CIOs and the Digital Transformation

Giorgio Bongiorno · Daniele Rizzo
Giovanni Vaia
Editors

CIOs and the Digital Transformation

A New Leadership Role

Editors
Giorgio Bongiorno
Comexposium
Paris
France

Giovanni Vaia
Department of Management
Ca' Foscari University of Venice
Venice
Italy

Daniele Rizzo
Autogrill S.p.A.
Milan
Italy

ISBN 978-3-319-31025-1   ISBN 978-3-319-31026-8 (eBook)
DOI 10.1007/978-3-319-31026-8

Library of Congress Control Number: 2017945711

© Springer International Publishing AG 2018
This work is subject to copyright. All rights are reserved by the Publisher, whether the whole or part of the material is concerned, specifically the rights of translation, reprinting, reuse of illustrations, recitation, broadcasting, reproduction on microfilms or in any other physical way, and transmission or information storage and retrieval, electronic adaptation, computer software, or by similar or dissimilar methodology now known or hereafter developed.
The use of general descriptive names, registered names, trademarks, service marks, etc. in this publication does not imply, even in the absence of a specific statement, that such names are exempt from the relevant protective laws and regulations and therefore free for general use.
The publisher, the authors and the editors are safe to assume that the advice and information in this book are believed to be true and accurate at the date of publication. Neither the publisher nor the authors or the editors give a warranty, express or implied, with respect to the material contained herein or for any errors or omissions that may have been made. The publisher remains neutral with regard to jurisdictional claims in published maps and institutional affiliations.

Printed on acid-free paper

This Springer imprint is published by Springer Nature
The registered company is Springer International Publishing AG
The registered company address is: Gewerbestrasse 11, 6330 Cham, Switzerland

Preface

It was 1995 when Nicholas Negroponte, in his book *"Being Digital"*, reminding the rapid migration from packaging logistic to the universally accessible transfer of data, predicted *"The change from atoms to bits being irrevocable and unstoppable"*.

These were among the first signals from the scientific community anticipating the dawn of a dramatic, albeit slow-coming, change.

Here we are, around 20 years later, witnessing our businesses and society finally becoming genuinely and exponentially digital, with an astonishing momentum similar to the that with which computers moved into our lives towards the end of the last millennium.

From then on, following Geoffrey Moore's *"Crossing the Chasm"*, *"Inside the Tornado"*, *"The Gorilla Game"* and *"Living on the Fault Line"*, the digital transformation became a necessary part of the charter of every company.

Today, this evolutionary process is progressing at a speed that was unthinkable until very recently. This is a time of aggressive and virulent transformations, which are affecting not only the Western world, but rather the entire world.

We will therefore briefly describe how this book came into being and how the goal of understanding the galloping digital transformation was developed.

Since 2001, *Finaki* [1] has organized a series of annual CIO community gatherings in Italy, with the intention of giving a *"voice"* and institutional recognition to the *demand* side of ICT in Italy.

Topics for each event are reviewed and selected by a Program Committee made of C-level participants belonging to leading Italian companies across several industries. *Finaki* is also the co-founder and platform provider for the European CIO Association, a not-for-profit international association which, since 2004, has brought together CIOs of the largest European businesses as well as CIO associations of Belgium, France, Germany, Hungary, Italy, the Netherlands, Sweden, the UK and Turkey.

---

[1]Finaki is a French company now owned by the *ComeXposium Group*, founded in 1989, pioneer and leader in high level information society focused events and conferences in Europe.

The sustaining pillars of the *Finaki* initiative are vendor independence, seasoned IT and Telecoms professionals combined with strong event management experience, and lastly, a limited number of selected participants, which guarantees the quality of exchanges.

This unique combination is the key to delivering networking opportunities, cross-feedback and open debates in a lively, informative environment to the senior IS executives community across Europe.

The topics covered by 2014, 2015 and 2016 events, respectively:

1. The Digital Revolution: a key opportunity to boost the country's economy
2. CIO: the Digital *Prometeus*. The Digital Connected Community: how to create and improve business in a digital integrated society
3. The dawn of the machine planet: how technology advances and ubiquitous digital transformation are redesigning our society

Through years of working with and analysing different situations and individual careers, we have identified sufficient professional material—joint reflections, considerations—to provide structured testimonials on how the *"demand side"* was both facing and hosting the ongoing digital transformation.

These collected reflections offer an overview of the perceived status of this change. We expect this contribution will be appreciated not only by the originating CIO community representatives but also by those of the general public who are interested in a deeper understanding of the ICT revolution and the related business transformation facts.

This book summarizes this opportunity, thusly providing a set of pragmatic approaches to address the controversial aspects of the *"digital transformation"*.

The following set of testimonials focus particularly on the effects of continuous technological transformations in a crucial component of the corporate digital organization: the ever-changing role of the CIO (Chief Information Officer).

Chapters are structured to host 2 major converging viewpoints: the managerial, represented by CIOs, and the academic, represented by university researchers.

This book starts by addressing the challenges faced by the CIO with a completely new role and continues with a description of the new digital governance needed to drive companies to a successful transformation journey.

We hope that these varied reflections will offer the reader a valuable image of new scenarios as well as some useful insights on the shifts that are currently in motion, calling for full digital integration.

Working, discussing and interacting with this particular set of people over sixteen years have raised many questions about how organizations in different sectors handle discontinuities and react to compelling change and about where to draw inspiration and insights from this revolutionary unstoppable and overwhelming transformation. Very often, we found different approaches and unbalanced mixtures of enthusiasm, successes, failures, disappointment and frustration. Also, we ultimately discovered that this revolution is not only a matter of technology, and therefore, we determined a need to reposition the human factors as central to the understanding of new contexts and correctly positioning their underlined uniqueness, ambiguity and uncertainty.

The set of testimonials found in this book should support the general need for a strong digital transformation, showing commercial and governmental organizations how effective a realistic governance effort can be in facilitating successful and lasting change.

CIOs, among other major top management players, are facing a period of unprecedented threats, innovation dilemmas and unexpected opportunities.

IT leaders are taking innovation initiatives and revamping or replacing traditional applications as critical priorities: this attitude appears to be based first on due respect for legacy, on the unmissable human aspects of this transformation and subsequently, as vital agents of change, on the willingness to correctly understand and interpret the unprecedented value and new digital business logic.

Investing in the future today, more than in the past, is of paramount importance in order to nurture the virtuous success cycle and to exploit the strategic potential of new emerging digital technologies.

Editors and authors are listed in alphabetical order. This order does not reflect the level of contribution and effort put in this editorial project.

Milan, Italy	Giorgio Bongiorno
Milan, Italy	Daniele Rizzo
Venice, Italy	Giovanni Vaia

Acknowledgements

We are particularly grateful to the various past Finaki Program Committee Presidents and Members, to the Vendors Representatives, to the Universities, to the Finaki Team and to all the colleagues that actively participated in our yearly meetings and contributed, with their field and corporate experience, a deeper understanding of one of the biggest, if not the biggest, transformations affecting our businesses, radically changing the way companies are responding to the market requirements and greatly influencing the design and the core values of our third millennium society.

Giorgio Bongiorno
Daniele Rizzo
Giovanni Vaia

Contents

CIOs and the Digital Transformation: A New Leadership Role 1
Giorgio Bongiorno, Daniele Rizzo and Giovanni Vaia

Human Being in the Digital World: Lessons from the Past for Future CIOs. 11
Francesco Varanini

The CIO and the Digital Challenge. 47
Daniele Rizzo

Future of the CIO: Towards an Enterpreneurial Role. 61
Carlo Alberto Carnevale Maffè

CIO's: Drivers or Followers of Digital Transformation?. 69
Giancarlo Capitani

CIOs at the Centre of a New Humanism 85
Bruno Demuru and Teodoro Katinis

The New Relations Among Things, Data and People: The Innovation Imperative. 107
Dario Castello, Gloria Gazzano and Giovanni Vaia

Digital Capabilities 121
Daria Arkhipova and Carlo Bozzoli

Designing the New Digital Innovation Environment. 147
Massimo Messina

Conceiving and Implementing the Digital Organization. 181
Mariano Corso, Gianluca Giovannetti, Luciano Guglielmi
and Giovanni Vaia

Digital IT Governance 205
William DeLone, Demetrio Migliorati and Giovanni Vaia

CIOs and the Digital Transformation: A New Leadership Role

Giorgio Bongiorno, Daniele Rizzo and Giovanni Vaia

Abstract Over the past few years, digital transformation and digital disruption have been widely addressed in both academic literature and business press. Still, these contributions notwithstanding, many IT executives are still struggling to find a go-to reference that would help them to orientate and fully exploit new digital opportunities through their transformational journey. This is partly due to the still limited collaboration between the two communities—academics and managers. Scholars are supposed to be focused on developing abstract models and theoretical representations of the new digital reality. Conversely, business and IT professionals seem to be primarily concerned with obtaining workable solutions to a concrete problem. In this book, we attempt to blend the two perspectives and explore new, fresh areas that are highly relevant for researchers and practitioners alike. To that end, each chapter of this book is written by scholars and co-authored with an IT executive, to bring together the rigorousness of academic research and the richness of practical experience in a single volume. This book is a collection of stories and experiences written jointly by academics and CIOs from the IT community, and is aimed particularly at managers and executives looking for inspiration to advance their digital journeys. The book opens with this introduction as an interview with Paolo Cinelli, Digital Business Manager at IKEA and one of the most influential figures in the international digital community. Paolo Cinelli talks about his experience with the digital transformation at IKEA, openly discusses the challenges he had to overcome along the way and shares his thoughts on what makes a CIO successful in today's digital world. This original contribution covers a wide range of experiences interlacing multicultural environments and gives an interesting

G. Bongiorno
Comexposium, Paris, France
e-mail: giorgiobongiorno359@gmail.com

D. Rizzo
Autogrill, Milan, Italy
e-mail: drizzo27@gmail.com

G. Vaia (✉)
Ca' Foscari University, Venice, Italy
e-mail: g.vaia@unive.it

perspective on the definition of Digital Transformation, the needed customer journey and the various challenges that this process is facing. Confronted with the various factors related to this unavoidable and ongoing change, a resilient manager should be able to recognize and seize all opportunities that every epochal disruption ultimately presents to existing organizations.

1 Introduction

Throughout history, new technologies have been disrupting well-established industry practices and replacing old, existing ways of doing business. The ongoing digital revolution is no exception—and while it may be perceived as a threat to some businesses, it presents unprecedented opportunities to others. The CIO could be at the center of this transformation and emerge as one of the leading actors and agents of this historical change.

New digital technologies affect both customer-side and internal processes of an organization, allowing a company to:

- Radically influence and shape management decisions
- Transform, often aggressively, the development of new products and services
- Find new ways to target customers by better understanding their needs
- Disrupt the existing business processes and models and create new sources of revenue
- Alter the existing industry structure and competitive landscape
- Enhance the quality of managerial decision-making
- Disrupt existing business processes and business models and
- Sensibly impact industry structures, customer experience and market strategic positioning.

What was initially perceived as a series of simple incremental product and process improvements is now fundamentally changing consumer behaviors, communication patterns, manufacturing and IT processes, as well as the types of products and services that can be introduced to the market.

Some skeptics may still think of "digital revolution" as another "buzzword" popularized by the media. In reality, the sheer magnitude of change related to digital technologies puts it on equal footing with the preceding industrial revolutions driven by stream-powered mechanization and electrically-powered assembly line production. What is different about digital technology today is the extent and speed with which it is pervading the lives of consumers, employees and organizations.

Back in 2000, the authors of the business bestseller "*The Cluetrain Manifesto*" (Levine et al. 2000)[1] predicted the effect that internet would have on well-established business practices that we take for granted today:

---

[1]Rick Levine, Christopher Locke, Doc Searls, David Weinberger (2000), The Cluetrain Manifesto: The End of Business as Usual, Basic Books, New York.

In just a few more years, the current homogenized "voice" of business, the sound of mission statements and brochures will seem as contrived and artificial as the language of the 18th century French court.

CIOs are center-stage to the digital transformation phenomenon. The very nature of a CIO's responsibilities undergoes substantial changes because in their new role, CIOs are expected to navigate through the multitude of opportunities, identify the right path for their business and drive adequate and timely change. To that end, digital transformation brings also a lot of uncertainty for CIOs: old-school approaches to managing IT no longer apply and new ones are emerging but not yet well-defined. At the same time, any transformational effort taken by a CIO will not simply affect IT function alone—it will have implications for an entire organization and for its customers and partners, too. Understandably, more often CIOs are cautious to make radical changes because the price of "getting it wrong" is sometimes too high.

The good news is that they are not alone in these challenges. Many companies that have managed to successfully "complete" (as if it can be ever be completed!) their digital transformation are very likely to have faced similar issues when they started. No transformation ever goes smoothly. The most successful digital transformations were carried out by companies that learned from their mistakes and pursued their goals relentlessly. Learning about their experiences can inspire new ideas as today's CIOs apply them to their own work.

One of the companies that has successfully integrated traditional and digital methods in its business is IKEA. IKEA was founded in Sweden in 1943 and is now the world's largest furniture retailer with a turnover of $36.3 billion. In 2015, IKEA reported 884 million store visits (7.7% increase compared to 2014), and 1.9 billion web visits (19% increase compared to 2014).

Paolo Cinelli, the former IKEA Group's CIO who currently holds the position of Digital Business Manager at the IKEA Franchisor Company (Inter IKEA), started the digital transformation at IKEA, and continues this work today. We had the opportunity to meet Paolo Cinelli and to get an exclusive interview with him. In the interview that follows, we asked Paolo Cinelli to share his experience on the topic of "digital transformation" and help us better understand how CIOs and IT professionals contribute to the process of value creation through digital. Furthermore, we wanted to explore his view on what could be a promising approach for organizations to manage digital transformation and to prepare for the challenges that might present themselves along the way. Below is our conversation with Paolo Cinelli.[2]

In a recent speech during the Finaki CIO Community Gathering you revealed some important principles for building an end to end "customer journey." Your statements portray you as candidate to steer the ongoing digital transformation process in your company.

---

[2]Paolo Cinelli gave permission to publish this interview.

How could you summarize the value of your international professional exposure?

An important part of my experience that inspires my work and attitude has to do with international exposure. I have had the privilege of working with different cultures, and in multicultural environments, developing a strong appreciation for their added value: ideas flow and flourish better, fostered by curiosity; interactions are more dynamic; discussions are richer thanks to multiple perspectives, etc. Unlocking such value requires an open mind and a positive interest in the differences around us, rather than an expectation that all must be the same. In my case, international exposure materialized in a number of assignments abroad, which have literally shaped myself and my family; but there are many ways to get the same exposure, so once again it is our self-awareness and willingness to change that make the difference.

My career goal was once to become a CIO, which I perceived as the "end station," but the reality is that once I got there and enjoyed the situation, my goal evolved. In fact, my energy decreased after a few CIO years, even touching a low point that triggered me to disclose my wish for a change. I was fortunate to be surrounded by a very humanistic company culture, and supported by great leaders who encouraged me to explore a next step outside my comfort zone. Still, had I kept this feeling of low energy, it probably would have resulted in low motivation, frustration and ultimately poor performance. Instead, expressing my desire for a change, combined with accelerating digitization trends and the trust of my colleagues, projected me into the exciting position of guiding the digital transformation of a very successful brand like IKEA. Although I could consider it a "dream job", I look at it with both passion and humbleness. What is a digital transformation? What do we mean with digital? How is digital different from ICT? Does it make sense to speak about a digital enterprise, or organization? This book tries to elaborate on those and many other questions through a multitude of views and reflections. One can agree with them or not, or partially agree, but the most important thing is to put the questions on the table and build possibilities to discuss them openly. I believe that is the spirit of this book, which therefore I read with high interest. And perhaps an ever important reflection is that the destination (e.g. becoming or remaining a CIO) is less enjoyable than the journey (as you go through it, you reach some milestones while the destination evolves into something even more attractive).

We believe that, conceptually, "digital transformation" boils down to a complex business transformation, of which technology innovations and people are the most important ingredients. What is a viable strategy you would suggest to successfully navigate the ongoing turmoil?

Although the word "strategy" is subject to many interpretations, a common assumption regarding strategic documents is that they have a well-defined target objective. But what if this assumption is unfounded? Is it possible to set a strategy (a "how") to achieve something that is not well defined? More than a year ago, in my company, we noticed the need for a new strategy or direction, and drafted a document called "IKEA digital business direction." Not surprisingly, it is a paper that has undergone multiple iterations, and is currently on version 11. What strikes me, however, is that one word on it was never questioned: "draft". It can be seen as a sign of hesitance, or insufficient buy-in, or the result of editorial difficulties, and so on. Actually, after almost a year I started to feel nervous about continuing to engage with stakeholders bearing a "draft" watermark on our key paper. Reflecting on it, though, I became more comfortable, especially when I observed that people are much more willing to engage on a draft strategy than on a sharply defined, signed off and static one. There is some logic in such behaviour: if an executive's role is to contribute at the strategic level to shape the future of the business, why should they spend time on strategies that are already fixed? Those are for the operational teams to execute! In other words, a term that might seem weak—draft—is actually the key to stimulate stakeholders' contributions, because they feel they can add their ideas, shape the strategy, and have their opinions considered; it leaves room for discussion, debate, involvement, etc. I can even argue that in some cases it might be smart to carry the draft status of a strategic document indefinitely, but I know that could raise skepticism, in particular due to the risk that without a signed off strategy, action is never taken. Let me then share what I noticed in that regard, at least in my organization. To my surprise, since I started sharing our draft with an increasing number of colleagues, I saw an acceleration of initiatives in the spirit of the (draft) digital direction. Initially, I even felt uncomfortable with that, and I realized that was due to the way we are raised in business. We grew up under the assumption that things evolve in a linear way: the business leaders outline expected scenarios, select the most likely one, create a strategy to address it, sign it off, deploy the strategy to convert it into business and action plans, which are in turn executed and monitored, so that we know when a strategy is effective. The reality, at least the current one, is different. We encourage initiative, and even entrepreneurship. Our companies are large and spread out, with plenty of clever leaders who want to make an impact. They enjoy being creative and see a mounting pressure for innovation around them. In such a context, sharing a draft strategy unleashes energy; people feel stimulated to try things out, and legitimated to possibly ask for forgiveness in case of failure, rather than for permission. At least that's what happened in our case, to the extent that a lot of suggested developments in our draft digital direction appear to be neither so farfetched nor innovative anymore, thus even questioning whether they should be mentioned in the strategy. Paradoxically, that calls for further revisions of the document, putting it in a status of "constant draft." Can this be a virtuous loop, advocating that in the contemporary, fast moving business world, linear planning is replaced by tentative strategies that, while constantly evolving, will never go beyond their "draft status"? Perhaps that goes too far or cannot be generalized, but, concerning the digital direction, I'm seriously tempted to believe in the power of "draft" and stay suspicious of "firm digital strategies". "Draft version" invites input, "final version" feels like "it's over, everything is already decided".

Provided that there is value in what you have defined as an everlasting "draft", what is your position on VUCA, applicable business strategies and scenario planning?

Today, many companies are struggling to look for additional resources, and they need to be extremely careful in choosing new technologies to deploy in ways that are most aligned to their businesses and strategic objectives. A quick search indicates that a strategy can be seen as a high level plan to achieve a goal. The search found the term to have entered the business language from military origins, and some sources add an element of mobilizing and optimizing resources, or a long term view. What is certainly common amongst all sources is the aim to a defined goal, and a plan or method to get there. I suppose that an implicit assumption is to know the starting point or current status from which you will arrive at the wanted one.

ICT strategies so far have followed suit. For example, the traditional approach to ICT architecture has been to define a target landscape and develop a roadmap to reach it from the current (legacy) baseline. As asked in chapter "Human Being in the Digital World: Lessons from the Past for Future CIOs", is that still a valid approach in the current environment? You have mentioned the VUCA situation we live and run business in: Volatility, Uncertainty, Complexity, Ambiguity. It can be argued that the entire universe, and therefore our planet and all life on it, have always been in such a situation, and always will be. But it's hard to deny that VUCA trends have recently accelerated for humanity and business. This situation makes it almost impossible to determine a target landscape, especially due to the disruptive speed of technology development. Can traditional architecture management methods still work in the current circumstances where a long term goal picture is undefined? My serious doubt about it leads me to further questions. What would be a better approach? Is adaptability superseding predictability? Consider an analogy based on a chess game, starting with unlimited time to make moves. The rules of a chess game have no ambiguity, and the goal of a checkmate is quite clear, but the complexity is so high that it's almost impossible to set a goal chessboard configuration (scenario). Still, players work very hard to explore the effect of possible moves and their ramifications, diving into predictable scenarios as deep as their thinking capacity allows. They visualize the chessboard landscape as a result of possible sequences of moves. Players with exceptional patience, concentration and logical reasoning capabilities (nowadays actually outplayed by super-computers) can imagine "target landscapes" that would be favorable to them, and steer the game in their direction. In case that doesn't sound challenging enough, let's introduce a timer, let's say 30 min per move (still quite a long time for professionals). Its effect can be twofold: on one hand, players can continue to use their same approach—identify a target chessboard landscape, and sequences of moves to get there—but speed up their reasoning; on the other hand, they can introduce a stricter selection of promising moves, likely based on probabilistic estimates or heuristics. Now, let's shorten the available time to, say, 10 min, and further down to 1 min. Does it still make sense, and is it at all feasible, to look for target landscapes? The complexity of the game hasn't really changed, but uncertainty has, because of the "limited visibility" one can develop into the future. We can still imagine an incredibly fast brain—or a super-processor—that can compute complex problems in a fraction of the time normally required; the challenge, however, can be stretched through further lead time reductions. Complexity can of course be increased, too, for example introducing an extra chessboard line and an extra column. The resulting mathematical challenge increases exponentially, and it would make the game volatile and uncertain if it was introduced in the middle of the game. Faster technological progress, additional complexity, mutating market rules and dynamics, shorter time to react, etc.: isn't this the business environment we all operate in? And given its characteristics, does it still make sense to try to define a target landscape? And what mindset do clever players choose in order to succeed in these situations? The research in this book indicates that shorter term scenario analyses and decision cycles, together with risk taking and an experimental attitude, as well as retaining flexibility to adapt to a VUCA context, are more suitable approaches than the traditional "set a long term goal and a roadmap to reach it" (or backward planning) one. In other words, we are called to invest in improving our ability of producing good heuristics, and getting comfortable in being guided by them with agility and speed. The good news is that even though VUCA triggers an urgent need to learn and adopt new mindsets and practices, those are useful paradigms in general, so once matured they become additional valuable options in the toolkit of ICT practices and organizations in general.

Your recipe for a successful business transformation seems to include three main ingredients: (1) staying optimistic, (2) focusing on opportunities rather than threats, (3) avoiding victimization. Is that so?

Do you believe that "resilience" is becoming the new emerging leadership paradigm?

An uncertain future, as well as any change in sight, can trigger different reactions. Are we intimidated? Excited? Curious? Skeptical? Something else entirely? The trick is to choose where we stand. I often see CIOs and other ICT leaders who fall back into a victim posture, which is of no help. When it comes to the digital transformation, what position do we choose? Do we look at the opportunities it offers, or at its threats? For example, I hear a lot of concerns about "pure-play digital retailers" (e.g. Amazon, Alibaba, etc.) seriously threatening brick & mortar ones. Of course that's a serious risk, but how do traditional retailers react? A few years ago, when we were stuck with the very same concern at my company, I heard a confident ICT leader asking a really good, energizing question: is it easier for an online retailer like Amazon to open 300 physical large stores, or for us to step up with our online channels? A simple question that paved the way for a much more positive mindset, and ultimately our omni-channel transformation. Resilience is also important, but not at the expense of adaptability and agility, otherwise there is a risk of becoming defensive of the past rather than willing to shape the future.

You have extensive experience working in organizations that were relatively digitally mature. How would you define the role of a CIO in these new scenarios? In your view, what makes a CIO successful?
How many question marks are scattered on the business transformation skyline?

We are talking about new digital organizations and enterprises, triggering important questions: to what extent should digital mastery be centralized in a company? How important is it to spread digital literacy across the organization? Or concentrate it in a few roles? And does that grant the CIO a unique/distinctive profile? Should there be a digital organization, and what is it exactly? Or are we evolving more and more towards eco-systems of competences? There probably isn't a one-size-fits-all answer, but that doesn't make those dilemmas less important, so what are the factors influencing the choice? In my opinion, digital technologies can be applied to so many different fields—product development, retail, manufacturing, customer engagement, logistics, etc.—that it's hard to think of a fully centralized competence pool, as it could become a bottleneck rather than an enabler. I would prefer to see "digital" as a resource that companies learn to utilize wherever suitable, in a mature way, just as they use different types of resources—financial, materials, processes, knowledge, etc. Still, I recognize the need for a transformational effort that addresses most of the above questions in the specific business context for the company and its culture; such transformation might require central stakeholders and dedicated resources, at least until it's well underway. Embedding digital competences in core business activities and development—hence in a decentralized way—can have a very positive impact on innovation, speed to market and suitability of new solutions; on the other hand, it creates complicated digital landscapes and the risk of diverging. How can CIOs play a proactive and stimulating role in this revolution, while preserving the information systems' integrity and efficiency? We should not see digital as being in contrast with information technology. Rather, they look to me like two sides of the same coin: the former much more visible and customer/user centric, the latter more internal and process efficiency driven.

What makes a CIO successful? What are the main leadership characteristics (distinctive and otherwise) that enable a CIO to be successful? One that stands out as crucial in my view is the ability to bridge different needs, ambitions and perspectives across the various areas of the business. The increasing level of digitalization increases the exposure of the CIO to virtually all the company's processes, and the importance of a bridge-builder grows with the complexity of organizations and their interactions. Further, the tension between control and innovation amplifies the importance of prioritization and allocation of company resources, so the CIO can facilitate at least the visibility of related opportunities, concurrent initiatives, interdependencies, etc. Such exposure can be perceived as overwhelming or as a privilege by the CIO; and the CIO's reaction is heavily influenced by her/his leadership: is the CIO leveraging the position for the advantage of the business, thus adding value, or just troubleshooting and juggling between multiple demands? Is the CIO showing challenges and opportunities in an engaging way that can unlock positive energy, or creating resistance and push back by focusing on constraints and limitations? Probably no single answer exists for all these questions, and the organizational culture plays a significant role in finding the right balance. I, however, believe the CIO should define his/her leadership profile in order to maximize his/her impact—whether it is "leading from behind", leveraging interdependence, boosting innovation, etc.—and align it with the executive team.

In your view, what can be considered the center of digital transformation? What is your professional advice to organizations that are undergoing digital transformation right now?

It is widely accepted that digital technology is driving a transformation across all industries. Its widespread adoption and exponential growth is disruptive to existing business models and its impact is accelerating. What's at the center of the digital revolution? Is there a center or are there multiple, concurrent pivot points conspiring into a "perfect storm"? Personally, although it would be easier to concentrate on a single factor, I am more inclined to look at the digital transformation as having multiple dimensions. The combined effect of concurrent trends amplifying each other is more important to understand and leverage than considering each of them in isolation. What is the combined effect of urbanization and digital? Is it compressing physical and time distance in such a way that it is generating a circular or shared economy, especially when adding the scarcity of natural resources? And what does a shared economy do to businesses that are heavily based on fixed assets? Those are just some examples of trends to be identified and assessed, but it's clear to me that the complexity of the situation requires looking at the reality with a holistic view and questioning established assumptions from the past. Once the context is understood well enough, the consequences can then be estimated by zooming into a single dimension, in our case "digital. "Is the digital transformation, that is currently challenging some of the ICT fundamental principles, as described in chapter "Human Being in the Digital World: Lessons from the Past for Future CIOs", so far universally accepted (reliability, security, stability, scalability, service continuity, etc.)? Are those principles still valid and necessary? Are organizations willing to sacrifice or compromise them in lieu of speed, agility, innovation, surprise, etc.? What competences should be involved in managing the transformation?

Your CIO leadership has a potentially very strong impact and influence within top management. ICT has always represented a sort of horizontal level of company spirit and business process management knowledge. Industrial, marketing and sales processes have been historically reinterpreted and digitalized with a substantial contribution from ICT. How does this relate to the topics of personal leadership and possible prejudice in the context of the ongoing "digital revolution"?

One of the expressions that irritates me the most is "ICT people." It is often used in a derogatory, even discriminatory fashion. The same is true when an equivalent expression is applied to any branch of the corporation. In my opinion, we are all people, regardless of our professional background, and like any stereotype, this kind of expression create silos and prejudice. So, we hear that "ICT people don't understand the business", "HR people are always…", etc. And we even take it seriously, thus accepting the assumption that ICT and "the business" are disjointed realities. Such a perception has existed for many years, but nothing can be further from the current situation, in which the convergence between the core business and its digital components is rapidly accelerating. Coming back to the people aspect, influencing how we personally and our communities are perceived starts with us internally: I cannot change others, I can only change myself. And yet, the impact of my own changes can definitely alter how I'm perceived by others (and ultimately influencing a change in them in some cases). Investing in our own leadership, customer understanding, professional mastery, etc., is the best way to earn respect and create a more accurate perception, regardless of what we do or are responsible for. Once I was fed up and raised my voice in a meeting with senior stakeholders when those derogatory expressions were used. I compared them to other categorization terms (e.g. "tall people", "southern people", etc.), and pointed out that they are in stark contrast with our own corporate values. It is important to draw a line on what is acceptable and what is not. In this case, the result is that I've never heard those expressions again in my presence. To reiterate, I was once hoping for others to change their rhetoric and opinions; not only was that not happening, but my frustration with them was growing. Instead, when I finally changed myself and reacted assertively, my colleagues changed, too, and my frustration decreased (I would like to say "vanished", but I cannot be sure that those expressions are not used when I'm not there, so I've decided to keep some healthy level of frustration so that I stay alert). What else shall I change in my leadership capabilities to increase the impact of my ideas, strategies, plans, etc.?

Thank you for taking the time to answer our questions. We are confident that our readers will greatly appreciate your contribution and will find your observations valuable.

Human Being in the Digital World: Lessons from the Past for Future CIOs

Francesco Varanini

Abstract Nowadays, it seems every company is racing to become more and more digital. But what does it really mean to "be digital"? For some, it is a matter of technology. For others, being digital is a new way to be in touch with customers. For still others, it is a completely new way of conducting business. None of these definitions is wrong per se, but each, by themselves, is only partially correct. The "digital disruption" forces us to consider not only business matters, such as methods of production, organizational operations, and money flow. The digital disruption changes every aspect of daily life for every citizen on the planet. This phenomenon might be challenging for some, such as the CIO, who were used to looking people as customers or company employees. Now, the CIO is called to equip everyone with certain tools and a habitable environment. To do so, the CIO's role must go beyond that of service provider and impartial observer. The CIO must redefine his own professional role by drawing from his personal experiences: only by reflecting on how he himself has changed and become a more digital human being will he be able to assist others in this process.

1 Introduction

One must first look back in time to better understand what the Digital Transformation entails and, therefore, be able to describe the new obligations and duties of the Chief Information Officer. A journey from the past to the future will illuminate our understanding of the present and help us to tackle new challenges with mindfulness.

The journey towards a new Digital World concerns all people. In this chapter, we observe the Digital World, not from an abstract, technical or scientific

F. Varanini (✉)
Department of Mathematics, Computer Science and Physics, Università degli Studi di Udine, Milan, Italy
e-mail: fvaranini@gmail.com
URL: http://www.francescovaranini.com

viewpoint, but rather from the perspective of the average person. Aside from being a professional, a technician, and a manager, the CIO is, first and foremost, a person. It is important to affirm this notion, given that the Digital Transformation opens the door towards a new world in which humans will coexist with 'autonomous machines,' equipped with Artificial Intelligence. As a result, the CIO is susceptible to being replaced by an algorithm.

Retracing the history of what we call *digital* is to retrace the history of all Information Technology as well as the history of the CIO's predecessor: the manager who provided the technological services necessary for business functions.

Digital is an adjective, but what is the noun? Let us consider the transition of the word digit, originally referring to a human finger or toe, to now being used to refer to an Arabic number symbol. We will look at the distinction between analogical machines and digital machines, and between analogical *codes* and digital *codes*.

Over the course of our journey, we will look at two elements—complementary, but different-of digital technologies: on one hand, we have infrastructures, or platforms. On the other hand, we have tools, devices, and applications that have now become crucial in the new digital world for allowing people to exercise their rights as free citizens, workers, and conscious consumers.

Throughout his existence, the figure we now call the CIO has always managed infrastructures and platforms. This trend is continuing in the new digital world. The difference is that now he must also manage the tools, devices and applications needed to guarantee everyone the ability to operate in their roles as citizens and workers. This new task requires a cultural change.

Looking too closely, or exclusively, at infrastructures and platforms, while ignoring one's personal experience, can be risky: on this path, it is easy for one to end up imagining a situation in which Designers create a world that is inhabited by citizens and workers, who are, in turn, reduced to the role of platform users, deprived of personal liberties and the ability to test their own creativity, accountability, and entrepreneurship.

Thinking only about infrastructures and platforms is not sufficient. The CIO must provide digital tools to all citizens and workers; these tools must be malleable and adaptable to the worker's needs. A good tool is co-constructed by the person who will use it.

It is important to consider the visions and personal narratives of the trailblazers who, in the 1940s and 1960s laid the foundations of the Digital Culture; these trailblazers include Vannevar Bush, JCR Licklider, Doug Engelbart, Ted Nelson. These visionary thinkers and technicians showed how technology can be a means for enlarging the consciousness of all people. CIOs can look to these thinkers and technicians as models: they taught us how to cultivate a vision, and of the importance of reconsidering business strategies in light of new opportunities offered by digital tools.

Only through experimenting can one become a digital citizen, only through first hand–hand experience with using tools can the CIO accompany citizens and workers in the transition to the Digital World.

2 A Humanistic Stance

Being Digital, an essay by Nicholas Negroponte, came out in 1995. This work marked an important turning point: the term *digital* left the technical lexicon of the Computer Science field and entered everyday language.

Digital: an adjective that distinguishes one type of machine from another.

During the 1930s and 1940s, two different types of computers existed: Digital Computers and Analog Computers. While the Analog Computer continuously measured the advancement of a process, the Digital Computer worked under a binary numeral system, in which data was converted into strings of 0 and 1. Analog Computing has not disappeared, but the machines that we know and use today are Digital Computers. These digital computers are based on the abstract idea of the Turing Machine, proposed by Alan Turing, and the digital machine architecture proposed by von Neumann.

Since the 1990s, however, *digital* is used not only in relation with machines, but also with people. People are invited to be *digital*. Negroponte wrote: until now, man has lived in a physical world, surrounded by material things. Now, we must prepare ourselves to live in a digital world.

Bits—the smallest unit of information, expressed in binary code—are rapidly replacing objects made of atoms. If we continue to pursue this lifestyle of *bits, not atoms*, Negroponte notes, the life of humans, who are becoming increasingly interconnected by computers, will never be the same.

Consequently, the role of Information and Communication Technology Director, more recently known as the Chief Information Officer, must evolve.

Traditionally, the CIO has worked with bits, data, and information. He considered humans as simply being users of machines and programs. The current digital scenario represents a new terrain on which people live and work. For this digital world to be more livable for people, adequate for their needs, and respectful of their rights, a new type of CIO is needed. It is no longer sufficient to have someone who works only with bits, data, and information; the CIO must be capable of hybridizing different fields of knowledge and action. Of course, he will need technical knowledge regarding the appropriate systems, infrastructure, hardware, and software. These hard skills, however, must be combined with soft skills in sociology, psychology, and ethnography.[1] Generally speaking, the CIO is called to take a humanistic stance to his technical position; this new position should be based on wisdom and mindfulness.[2] The CIO is, after all, a human like all others.

---

[1]Francesco Varanini, "Il ricercatore debole, o La restituzione poetica", in Gianluca Bocchi and Francesco Varanini, *Le vie della formazione. Creatività, innovazione, complessità*, Guerini e Associati, Milano, 2013, pp. 57–69.

[2]Francesco Varanini, "Complexity in Projects: A Humanistic View", in Francesco Varanini and Walter Ginevri, *Projects and Complexity*, CRC Press, Boca Ratón, 2012.

3 What Is Digital

Being Digital is a collection of columns written by Negroponte for the monthly publication *Wired*, a magazine he himself helped to found in 1993. *Wired* predicted the imminent arrival of a Digital Revolution. The magazine has its headquarters in San Francisco, just a few miles from Silicon Valley and Stanford University, in the heart of an area in which, currently, a new technology, a new culture, and a new economy are being born.

Being Digital was immediately distributed on a global scale. In 1995, it was translated into German, French, Italian and Spanish. The Japanese version came in 2001.[3]

By 1967, around the age of 24, the young Negroponte, son of a rich, Greek, ship owner, was already managing the Architecture Machine Group, a laboratory and think tank dedicated to the study of human/computer interaction, at the Massachusetts Institute of Technology. In 1985, to further this research, Negroponte founded the Media Lab at MIT, with the help of Jerome B. Wiesner.

In both English and Latin, the word "media" is the plural of medium. *Media*, treated as singular or plural, can mean 'main means of mass communication (broadcasting, publishing, and the Internet) regarded collectively'. But this definition is not what Negroponte was referring to.

Negroponte favored the concept used by Canadian philosopher and semiologist Marshall McLuhan who, in 1964, published *Understanding Media*.[4] For McLuhan, *medium* is synonymous with technology. A medium is "any new technology." One famous example is the lightbulb: "a light bulb creates an environment by its mere presence." It is a medium, or a technology, and, like any other medium, has a social effect. "The 'message' of any medium or technology is the change or scale or pace or pattern that it introduces into human affairs."[5] The lightbulb allows humans to transform the dark of night into a livable space. Similarly, it changed man's way of experiencing each new medium that followed: the train, the automobile, the radio, and the television.

---

[3]Nicholas Negroponte, *Being Digital*, Knopf, New York 1995. German translation: *Total digital: die Welt zwischen 0 und 1 oder die Zukunft der Kommunikation*, Bertelsmann, München, 1995. French translation: *L'homme numérique*, Laffont, Paris, 1996. Italian translation: *Essere digitali*, Sperling and Kupfer, Milano, 1995. Spanish edition: Ser digital: Editorial Atlántida, Buenos Aires, 1995. Chinese translation: ビーイング・デジタル - ビットの時代 新装版 [Digital Revolution], New Taipei (Taiwan) 1997. Japanese edition: [Being Digital: The Bit Era], Asukī, Tōkyō, 2001. Audiobook: Nicholas Negroponte; Penn Jillette, Being Digital, Random House Audiobooks, New York, 1994.

[4]Marshall McLuhan, *Understanding Media. The extension of Man*, MacGrow Hill, New York, 1964.

[5]Marshall McLuhan, Understanding Media, p. 8.

Even more significant is the change in the human environment caused by the pervasive presence of computers, each connected to each other and to humans through interfaces.

We have become used to hearing the word *digital*. As a result, the French translation of Negroponte's title—*L'homme numérique*-baffles us; it surprises us and raises some questions. We are not used to substituting *digital* with *numeric*. It is well known that the French are always looking to translate English words—hence *software* becomes *logiciel*. But while *logiciel* does not fully capture the sense of the original English word, *numérique* seems to be a perfectly correct translation, which draws on the hidden history of *digital*.

Let us discuss the Digital Computer; this machine functions by means of codes expressed in chains of 0 and 1. In reality, however, the word *digit* contains no reference to binary numeration.

In Latin, *digitus* means 'finger.' From *digits* comes the Italian *dito*, the Spanish *dedo*, and the French *doigt*. *Digitalis* means 'from a finger', 'having the dimensions of a finger', and 'having the shape of a finger.' A plant that has a shape similar to a finger also takes this name.

The origin lies in the Indo-European root *deik*, which meanings 'to indicate', 'to show', or 'to point out.' This root appears in other language families as well, including the Sanskrit dic-, the ancient Greek *deiknynai* and the German *zeigen*, all of which are verbs meaning 'to show.' What's more, the Latin verb *dicere*, still from the same root, means 'to say' or 'to speak.'

Perhaps the word *toe* comes from the same root as *deik* as well.

Finger comes from the early Germanic *fingraz*, probably from the Indo-European root *penkwe*, meaning 'five.' Five, as in the number of fingers on a hand. Using the fingers on our hands—with ten, five, two—man learned to count.

By using our fingers, humans are able to distinguish 'I' from 'you' and from other people. We can also indicate the various events and phenomena that surround us.

And finally, with our hands, we create utensils, tools, and instruments.

4 An Idea of the Technique

The full title of McLuhan's essay is *Understanding Media: The Extensions of Man*. Each *technique* is intended as an extension of the human mind and body. Here, I am using the word *technique* and not *technology*, because technique is more vast, historical, and philosophical, while technology is a new word, essentially American, associated with the founding of the Massachusetts Institute of Technology in 1865. Technology is a calque from the ancient Greek *technologia*, 'work related to an art.' Technique, a French word integrated into English, harkens more directly back to the ancient Greek *téchnē*, 'art', 'skill', or 'craft in work'. This word originates from the Indo-European root *tek*, 'to create.' From the dawn of time, man has been able to think and work because he has a mind and hands. We are reminded of this notion with the motto of the Massachusetts Institute of Technology: "Mens et Manus," mind and hand. Man, through thinking and working, learns through experience and, ultimately, comes to know. In Latin: *experio*: 'I try.' As such, through trial and error, man creates extensions of his own body, mind, tools, and *artifacts*. *Artifact* means simply 'handmade.' The Latin word *ars* corresponds to the Greek *téchnē*: it means technique, but also *art*—'expression or application of human creative skills and imagination.' This word originates from the Indo-European root *are*, meaning 'to adapt,' from which comes *art* and also *arm*, referring both to the upper limbs of the human body and weapons.

Thus, *technique* is the art of constructing tools, means, media, and appliances—an art form that has been well known by all men since prehistoric times. *Technique*, or *technology*, is not simply a type of practical or applied science, but rather a term that can be used to describe all *voluntary extensions of natural processes*. As such, if breathing is a necessary human function, then the ability to breathe underwater would be, in a sense, the result of technology.

In 1964, the same year as *Understanding Media*, published by McGraw-Hill in New York, the first volume of André Leroi-Gourhan's essay was published by Albin Michel in Paris, *Le geste et la parole*.[6] The volume was entitled *Technics and Language* (in the original French language edition, the author used the word *technique*. The English language translator sometimes used *technics*, other times *technique*, and other times *technology*).

Leroi-Gourhan is a paleontologist, archaeologist, and anthropologist. He harbors a special interest in technology and offers readers the opportunity to experience the dawn of humanity in the 1960s, the period in which the culture now known as digital established itself.

In *Understanding Media*, McLuhan offers the same experience to readers, though the historical aspect is limited. Leroi-Gourhan could easily use McLuhan's subtitle, *The Extensions of Man*. Indeed, he illustrates the meaning of this subtitle better than McLuhan did.

Leroi-Gourhan begins his story at the moment in which humans started to stand upright, differentiating themselves from other animals. The development of the front part of the head allowed for the development of the brain, along with human intelligence; the development of arms and hands allowed people to collect fruit and use stones and wood as tools. These tools became a sort of artificial limbs, *extensions of man*.

Leroi-Gourhan makes note of this evolution, characterized by man progressing from using his hands to using tools created by hand. The author notes also how man progressively transfers his intelligence to things he creates.

Leroi-Gourhan stopped writing around the mid-1960s, but not because the technology presiding over the construction and use of computers (machines capable

---

[6] André Leroi-Gourhan, *Le geste et la parole*, 2 vols. (Paris: Albin Michel, vol I Technique et langage, 1964–vol II *La mémoire et les rythmes*, 1965. English translation: *Gesture and Speech* (Cambridge, Massachusetts & London: MIT Press, 1993).

of replacing people in the workplace) was fully mature. Rather, these years were also the period in which he invested much of himself in his research on the development of Artificial Intelligence.

Technique is therefore a process powered by man, but that is also leading to man's marginalization. The tool, which was initially a man-made extension of man, is progressively separating itself from man, with no signs of returning. Hence, we might say that we are heading towards a future in which machines are becoming increasingly independent from man, even being constructed by other machines.[7]

Current trends are increasingly pushing us to believe in digital intelligences that are different from human intelligences. Starting in the 1940s, various expressions were used to label machines that defined for themselves how they would work. The following expressions were often heard: Cybernetics, auto-regulation, Artificial Intelligence, Machine Learning, Self-Managed, Self-Operating, Self-Repairing, Self-Sustaining, and Self-Driving Machines. More concisely, all of these expressions relate to an *autonomous* nature, a word coming from the Greek *autos*, 'self,' *nomos*, 'law.'

The role of ICT Director was born in the 1960s, when autonomous Non-Human Systems were nothing more than a project or a concept from Science Fiction. The CIO, a role which evolved from the ICT Director, came about in the new millennium. This person found himself faced with the task of having to manage increasingly autonomous systems.[8]

Against this backdrop, one must wonder if it is unrealistic or too idealistic to look at the Digital Revolution as a time in history that created new liberties for people. Instead, maybe the Digital Revolution was just a triumph for autonomous machines. Maybe it would be better not to talk about a new human environment. Maybe it is better to say that the CIO works for autonomous machines. Maybe it is better to say that the human CIO is preparing the terrain for a revolution that would lead to his own substitution by a software CIO or algorithm CIO.

5   Autonomous Machines and Autonomous Human Beings

"Quite soon, the world's information infrastructure is going to reach a level of scale and complexity that will force scientists and engineers to think about it in an entirely new way",[9] writes Mark Burgess, as an opening to *In Search of Certainty*. Burgess is the designer of CFEngine (Configuration Engine, a software framework that automates the configuration and maintenance of infrastructure). If this is true for scientists and engineers, it is even truer for CIOs. One must think in a new way.

---

[7]Francesco Varanini, *Macchine per pensare*, cit., 2016, pp. 235–266.
[8]Francesco Varanini, *Macchine per pensare*, cit., 2016, pp. 235–266.
[9]Mark, Burgess, *In Search of Certainty. The Science of Our Information Infrastructure*, O'Reilly, Sebastopol, CA, 2015a. Second edition, p. 1. (First edition 2013).

"The myth of the machine, that does exactly what we tell it, has come to an end." We now find ourselves obligated to manage autonomous machines. As such, the task is actually to manage one complex, autonomous machine, an information infrastructure. A business's information infrastructure cannot be seen as being isolated, as it is part of the world's information infrastructure. It is no longer a collection of separate parts that can be managed individually, but rather a collection of weakly coupled elements that interact and cooperate. This change is the defining feature of the revolution, the disruption: we have been thrown into a "faster, denser world of communication, a world where choice, variety, and indeterminism rule."

Talking about digital risks being limiting or misleading. *Digital* exists because it relies on the world's information infrastructure. CIOs are asked not only to manage within their own realms, but also in this global structure.

Burgess knows exactly what this entails. He is the designer of CFEngine, an agent/software robot and a high-level policy language for building expert systems aimed at the configuration and maintenance of large-scale computer systems, including the unified management of servers, desktops, consumer and industrial devices, embedded networked devices, mobile smartphones, and tablet computers.

"We suffer sometimes from the hubris of believing that control is a matter of applying sufficient force, or a sufficiently detailed set of instructions." Using the co-presence and interaction between human and non-human (software, algorithms) agents, Burgess shows how management models founded on the traditional "command and control" approach, in which the central authority orders agents to behave a certain way, is increasingly inadequate when faced with the information infrastructure.

By observing the realms of physics and biology, one can understand how uncertainty is an inescapable fact of life. From this observation emerges a new approach for governing the infrastructure, in other words, for governing the interactions between humans and non-humans.[10]

Burgess spoke of the *promise theory*: autonomous agents declare their own behaviors in the form of promises. The trust established between agents is the fruit of promises made and kept in the past. The behavior of the system emerges from interactions. It is a bottom-up, constructionist view of the world.[11] It is an important idea for the Digital CIO, both in terms of managing the machines and in terms of relations between people.

The machines—the singular parts of the information infrastructure-are no longer seen as a type of hardware whose function can be completely understood, or as software whose coding and documentation can be understood. Now, machines are accepted as agents of whom the behavior is observed in action.

---

[10]Federico Cabitza and Francesco Varanini, "Going beyond the System in Systems Thinking: the cybork", cit., 2017.

[11]Mark, Burgess, *Thinking in Promises: Designing Systems for Cooperation*, O'Reilly, Sebastopol, CA, 2015b.

Humans are, in turn, accepted as having a degree of liberty in their actions. They are no longer reduced to predefined roles: data entry specialists, programmers, users…In principle, people are capable of any type of behavior. The behaviors, done over time, define the ever-evolving profile of a person.

The environment, in this perspective, is not the result of a pre-defined image, or the fruit of some project. The environment is the terrain occupied by the agents through their explorations. Furthermore, in this perspective, it would seem improper and reductive to speak of a human environment. We are speaking instead of an environment adapted to humans and machines alike.

One does not need to search for a new environment for humans in the Digital Revolution, but rather a terrain where people—autonomous agents in the context of a complex system—coexist with non-human agents, continuously undergoing reciprocal adaptations. Burgess writes "What we seek, in pursuing human-computer relations is a balance between the dynamical stability [typical of machines] and semantic creativity [typical of human beings]. It must allow the business of society to prosper in a predictable and trustworthy way."[12]

6 A Perennial Gale of Creative Destruction

Burgess does not observe the world from an outsider's perspective. He cannot imagine trusting the governing of the infrastructure to a software CIO or an algorithm CIO. Burgess's view remains humanistic and does not recognize the potential autonomy of machines. "Technology and machinery exist for the benefit of humans, and we should not forget that".[13] The actions of the non-human agents are considered in terms of the advantages they provide to humans: "a way to take pointless and bothersome relationships away from humans, freeing them to think about issues more worthy of human dignity".[14]

Burgess is a technician, close to CIO, who enlarges his viewpoint to include the ethical and social implications of the ubiquity and pervasiveness of digital infrastructures. Still, he maintains the infrastructure's role as an *underlying layer*, a *lower layer*, on which other, social agents, managers, and political leaders outline social structures, as a *higher layer*, an *overarching layer*.

If one wishes to look, without illusion, at the scenario created by the Digital Revolution, one must go beyond Burgess and read a more radical view of the situation we are currently witnessing.

Leroi-Gourhan and McLuhan remind us that *technique* is an *extension of man*, born of man. Still, it is evolving in a way that leads to autonomous machines. The *information infrastructure* is the terrain on which humans play out their entire lives.

---

[12]Mark Burgess, *In Search of Certainty*, p. 313.
[13]Mark Burgess, *In Search of Certainty*, p. 397.
[14]Mark Burgess, *In Search of Certainty*, pp. 296–297.

But it is also the terrain on which non-human actors are present. For this reason, it is not appropriate to limit one's view of the information infrastructure as being an *underlying layer*. The technician's perspective is insufficient. A more complex view is necessary, one that takes into account ethical, social, and political implications of the new scenario.[15]

Benjamin H. Bratton, in *The Stack*, starts with a very clear statement about this phenomenon. "The model does not put technology 'inside' a 'society', but sees a technological totality as the armature of the social itself."[16] We must not look at the infrastructure proposed by some group of technicians, but rather at "an accidental megastructure [...] that is not only a kind of planetary-scale computing system; it is also a new architecture for how we divide up the world into sovereign spaces."[17]

That which Burgess calls infrastructure can more precisely be referred to as a "multilayered structure of software protocol stacks in which network technologies operate within a modular and interdependent order".[18] But what is more deserving of attention is how this infrastructure "is changing not only how governments govern, but also what governance even is in the first place".[19] These changes are the consequences of the Digital Revolution.

Here, we are not talking about a technological *underlying layer*, nor an enabling infrastructure, but rather a *megastructure*. This concept of *structure* requires further reflection.

We can see, in this word, the Indo-European root *ster*, meaning 'to spread out.' We can also see the ancient Greek root *stratos*, 'army deployed,' and *strategôs*, 'army chief,' the latter also being the root for the word *strategy*.

The Latin verb *struere*, meaning 'arrange one layer over another,' comes from the root *ster*. From here comes the Latin *construo*, origin of the English *construction*, and the Latin *destruo*, from which comes the English *destruction*. From here also comes the Latin word for an abstract concept: *structura*. From the verb *struere* comes also the Latin *stratum*, meaning 'layer', origin of the English word *street*.

Therefore, the *structure* is a continuous attempt, a continuous piling of layers, one over the other. With each new layer, one must decide whether to add more, remove older layers, or change the position of existing layers.

Let's look now at the meaning of infra. The Latin *infra* comes from the same Indo-European root as *under*. Infra is a contraction of infera, which comes from an even lower layer: the Latin *inferus*—from which comes *inferno*, 'hell'—meaning *lower*.

---

[15]Francesco Varanini, *Macchine per pensare. L'informatica come prosecuzione della filosofia con altri mezzi (Trattato di Informatica Umanistica*, vol. 1), Guerini e Associati, Milano, 2016.

[16]Benjamin H. Bratton, *The Stack: On Software and Sovereignty*, The MIT Press, Cambridge, Ma. 2015, Preface, p. xviii.

[17]Benjamin H. Bratton, *The Stack*, Preface, p. xviii.

[18]Benjamin H. Bratton, *The Stack*, Preface, p. xviii.

[19]Benjamin H. Bratton, *The Stack*, Preface, p. xvii.

Thus, we have reason to use *infrastructure* to mean the lowest level of a *structure*, which is a stack, a temporary overlap of layers.

Burgess identified a series of layers that he deems the most significant.

Earth

"There is no planetary-scale computation without a planet".[20] "Planetary-scale computation needs smart grids to grow, and for smart grids to grow, they need more ubiquitous computation".[21]

Cloud

"The Cloud layer is also a geopolitical machine, erasing some geographies and producing others, forming and destabilizing territories in competitive measure."[22] "The geopolitics of the Cloud are everywhere and want everything: the platform wars between Google, Facebook, Apple, and Amazon".[23] We cannot forget that the cloud also hosts the Deep Web. Economic transitions conducted without regard to Earth laws and bit-coin financial operations are nothing but examples of the digital life that exists in the Cloud.

The Cloud is the setting of new form of politics and geography. We must imagine that even the businesses for which we work are moving towards this new territory: they now longer exist exclusively in the Earth, but also in the Cloud. If we must move towards the Cloud, we must also have a manager—the CIO—who is able to accompany the business in this transition, towards this new territory.

City

The smart city, a place inhabited by humans, a place for settlement and for mobility, has been remapped to more closely resemble the platforms present in the Cloud.

Address

The digital revolution imposes "the addressing of every 'thing' therein that might compute or be computed."[24] Each thing, each human, and each machine is described with a tag, a synthetic address, in order to allow for connections and transfers.

Interface

"An interface is any point of contact between two complex systems that governs the conditions of exchange between those systems".[25]

---

[20] Benjamin H. Bratton, *The Stack*, p. 75.
[21] Benjamin H. Bratton, *The Stack*, p. 93.
[22] Benjamin H. Bratton, *The Stack*, p. 110.
[23] Benjamin H. Bratton, *The Stack*, p. 110.
[24] Benjamin H. Bratton, *The Stack*, p. 191.
[25] Benjamin H. Bratton, *The Stack*, p. 220.

In Latin *inter* means 'between.' *Faciem* is an abstraction from the verb *facere*, 'to do.' That which is done has a form, an aspect: therefore, *a faciem*, a face.

To understand the meaning of *interface*, we must first consider *surface*. In Latin, it is *superficiem*: the prefix *super* 'above', 'over', 'on the top', is placed before *faciem*.

Super is the opposite of *sub*, 'under', 'beneath.' Behind *super* and *sub*, one notion is present: the stasis on a surface and the vertical movement from low to high. We notice the same exact concept present in the word *structure*.

Users

In Latin *utens*, from which comes the English 'user,' is 'qui utitur aliqua re', 'one who uses a thing.' The Latin verb *uti* translates exactly to 'use.' The user is thus the person who uses a tool or device. The user is a human being who uses a digital platform. A digital platform—for example, Facebook used as a company's Employees' Portal—has its own rules and laws. Someone, such as the CIO, who builds and manages these platforms has the difficult role of legislator. We can look with hope to the establishment of a future digital citizenship, but there is a difference between the citizen and the user.

The German philosopher and student of Matin Heidegger, Hannat Arent, speaks quite clearly about the human condition as an attempt to fill spaces of citizenship. The citizen is an active player: in Latin *agens*: a person who acts, powerful. The user is a passive player: in Latin *patiens*, a person receiving care.[26]

Bratton offers us a series of layers described one by one. But even he warns us to be on alert; his description is only an attempt, a proposed breakdown or system for distinguishing elements. Layers are, in reality, intermixed and never separable. What is important is the new complex image of the world: a stack, a structure, an infrastructure. This is the environment in which humans live after the digital disruption.

Burgess, with his limited way of thinking about the specialized technician in a technological *layer*, forced us to think of the CIO as a manager of staff. A manager that manages the *underlying layer*, *lower layer*. A manager that offers support to other managers who, at a higher level, take charge of strategic and political decisions.

Bratton, by describing a structure as a stack of layers, illustrates the vast responsibility of the CIO. The CIO is not only a technician. He is the manager who, more than any other, is capable of understanding and accepting the new terrain on which people and machines coexist. The CIO is, more than any other professional figure, able to understand the complexity of the Layered Infrastructure in which people and machines interact. No one is more capable than the CIO of understanding the implicit difficulties inherent to double management: management of

---

[26]Hannah Arendt, *The Human Condition*, 2d edition, The University of Chicago Press, Chicago, 1988, p. 175. Vedi anche Chapter III. Labor, 17. A Consumer's Society, p. 126 and following. First published in 1958.

people and machines. Only the CIO has an accurate perception of how the *underlying layer, lower layer* does not exist: the pile of layers, as Bratton shows, according to the perennial gale of Creative Destruction: layers are constantly changing and cannot be managed separately. The CIO of the future will not only be coauthor of each strategy and each company policy, but he will also have something to teach political leaders and social reformers.

7 A Divine Escape

On Thursday February 16, 2017, Mark Zuckerberg, founder of Facebook, published a long post on Facebook entitled *Building Global Community*.[27] It was more than a speech coming from a chief of state, it was an encyclopedia, a pastoral letter from the pope of the universal church.

In the text, the word infrastructure appears in the fifth line, and then again twenty-four more times. "In times like these, the most important thing we at Facebook can do is develop the social infrastructure to give people the power to build a global community that works for all of us".

Infrastructure is defined with demanding adjectives: social, meaningful, global. Equally demanding are the adjectives used to define the Communities in which people are claimed to deserve to live: supportive, safe, informed, civically-engaged, inclusive. One immediately understands that Zuckerberg considers Infrastructure as being synonymous with Community.

Mark signed the letter with just his first name, like a real pope.

Zuckerberg also offers his thanks: "Thank you for being part of this community." We can understand this to mean: Thank you for living in Facebook, The Infrastructure. A new digital environment is proposed—or rather, imposed—on people.

Zuckerberg, in an address to all people, calls us to participate in the construction of the Infrastructure, "the world we want for generations to come." He seems to forget that Facebook already exists as a platform and that none of us participated in its construction. The citizen is reduced to a user. Loading materials onto Facebook does not constitute participation. Our knowledge and abilities are limited to the pre-defined format and contents of Facebook. In a true situation of freedom, people could choose the way in which they express themselves. But Facebook imposes rules and forms. In the confines of such a pre-established Infrastructure, our knowledge is reduced to content, the thing that is contained, cooped, forced. This problem concerns all software developers and CIOs.

We must wonder if we are inhabitants of the world, like all human beings, or if we are designers of the world in which other human beings will have to live.

---

[27]Mark Zuckerberg, *Building Global Community*, https://www.facebook.com/notes/mark-zuckerberg/building-global-community/10103508221158471/.

The entire history of the development of software is characterized by this question. Such a question implies series ethical doubts to programmers who limit themselves to developing a single software, a single application aimed at helping a person complete a single activity. The question becomes even more serious in light of the digital disruption, a period in which we are effectively capable of constructing the entire world in which people will live, as in film Truman Show.

Understanding Computers and Cognition: A New Foundation for Design, published in 1986, is a book that signaled a changing point. The authors are Terry Winograd and Fernando Flores.[28]

At a young age, Flores, of Chile, became the Minister of Economic Development and the Minister of Finance in the government of Unidad Popular. After the overthrow of the government in 1973, he built a new career as a Computer Science researcher at Stanford University. There, he met Terry Winograd, professor of Computer Science, already known for his work in natural language systems and human/computer interaction. In 1967, the need for consideration to ethics led Winograd to found CPSR, Computer Professionals for Social Responsibility.

In *Understanding Computers and Cognition*, Winograd and Flores adopt a conscious *humanistic stance*, or, better, given the explicit reference to the German philosopher Heidegger, a conscious *phenomenological stance*. The Greek *phainómenon*, which is 'that which shows itself', 'that which appears—to man' while man lives in his own world and learns through experience. The behaviors of humans who, through work or play, come in contact with computers, do not intersect much with the behaviors of a computer scientist or an Information Systems specialist who works with data and information. The computer scientist, through his training and education, bases his behaviors on logical deduction and conscious reflection. With people who interact with devices on a more personal level, however, individual interpretation and intuitive reasoning play a central role. Winograd and Flores sought to layout the basis for a programming system aimed not at imposing a predefined type of behavior on people, but rather a system that would take into account the behaviors of people.

It was not by chance that Winograd and Flores referenced two important concepts from Heidegger's principal work, "*Being and Time*".[29]

Throwness (Geworfenheit):[30] our Being-in-the-world is being thrown into the world. The world is not a comfortable, protective, welcoming place. We are living in a wasteland. The *Throwness* imposes responsibilities on us, but also offers us possibilities.

---

[28]Terry Winograd and Fernando Flores, *Understanding Computers and Cognition: A New Foundation for Design*, Addison-Wesley, Reading, Mass, 1986.

[29]Martin Heidegger, *Sein und Zeit*, M. Nyemeyer, Halle an der Saale, 1927. English edition: *Being and Time*, translated by Joan Stambaugh, University of New York Press, Albany, NY, 1996.

[30]Martin Heidegger, Sein und Zeit, § 29, § 31, § 38, § 42, § 68B. Winograd and Flores, *Understanding Computers*, § 3.3, p. 33 and following.

Readiness-to-hand (*Zuhandenheit*):[31] "The hammering itself uncovers the specific 'manipulability' of the hammer", writes Heidegger. "The kind of Being which equipment possesses—in which it manifests itself in its own right—we call 'readiness-to-hand'".[32] We have seen, according to Leroi-Gourhan's reasoning, how certain tools tend to become machines that are separate from people, *things* that are distant from people, in opposition to humans. One example is the computer, which is moving towards complete autonomy, and even Artificial Intelligence. *Ready to hand* is the hammer with which people learn by experience. The human experience manifests itself in the use of utensils. This word derives from the Latin verb *uti* 'to use,' *utens*, 'user', and *utensilis*, 'tool.' Winograd and Flores force themselves to imagine digital machines not as autonomous machines, but rather as *utensilia*, tools, always ready to hand.[33] Here, we can think about personal computers, tablets, and smartphones.

Heidegger's *Readiness-to-hand*, through Winograd, Flores and other researchers of their era, such as Donald Norman,[34] was the source of research dedicated to Human Interface (HI), Human-Computer Interaction (HCI), Usability, User Interface (UI), User Experience, (UX).

The meaning of what we do while using tools depends on the context and the situation in which we are thrown.

Heidegger distinguishes between *Sein*—which translates directly to the English *being*—and *Dasein*—which translates to *being there* or *presence*. Given the importance that Heidegger attributed to *Dasein*, the English translators of *Sein und Zeit*, preferred not to translate it, leaving instead *da-sein* in the English text. But Heidegger is less obscure than one might initially think. Our *being-in-the-world* cannot be programmed nor predicted, nor described by an outside source. *Dasein* is 'to feel in a certain situation,' it is the 'how things are going.' The German *da* refers to the *state of mind, understand*, and can refer to both *there* and *here*.

The *headline* of an AT&T publicity campaign, released in the period in which Negroponte was pushing *Being Digital*, was: *Be there here*. The publicity campaign was referring to a non-place, a virtual space, the cyberspace that two people share during a telephone conversation. That which was true for the telephone conversation then is even truer for us now on the various platforms and digital infrastructures. We have been thrown into an unknown world. We are constantly faced with *being* in this world.

---

[31]Martin Heidegger, *Sein und Zeit*, § 15, § 18, § 22, § 69A. Winograd and Flores, *Understanding Computers*, § 3.4, p. 36 and following.

[32]Martin Heidegger, *Sein und Zeit*, § 15. English edition, *cit.* p. 98.

[33]Francesco Varanini, "Complexity in Projects: A Humanistic View", cit, 2012, pp. 52–55. Francesco Varanini, *Macchine per pensare*, cit., 2016, pp. 261–266.

[34]Donald Norman and Stephen Draper (eds.), *User Centered System Design: New Perspectives on Human-Computer Interaction,* L. Erlbaum Associates, Hillsdale, N.J., 1986. Donald Norman, *The Psychology of Everyday Things*, Basic Books, New York, 1988; with new title: *The Design of Everyday Things*, Currency Doubleday, New York, 1990.

The humanistic or phenomenological stance is living *Dasein*. The environment —in German *Umwelt*—is the surroundings, and the world around us here and now. The environment to which we need to adapt surely includes the digital world, the infrastructure. But using Facebook, or, more generally, using the World Wide Web, does not mean, despite what Zuckerberg might want, reducing ourselves to living in a world designed for us by Zuckerberg, or by a CIO. Winograd and Flores question how people can live *Dasein*, along with a *readiness-to-hand* machine.

That is where the shoe pinches. We imagine that Winograd and Flores are interested by the Dasein, the state of mind, understanding people who are thrown into the world equipped only with digital tools. But that is not the case. The first two parts of *Understanding Computers* are simply a preparation for the third part, entitled *Design*, which reinforces the book's subtitle: *A New Foundation for Design*.[35] Heidegger tells us that before being technicians, programmers, computer scientists, CIOs, or managers, we are people, human beings who, like all other humans, are thrown into a world that we must come to understand.

Heidegger tells us that just by assuming a humanistic stance, we can eventually come to be good technicians, programmers, computer scientists, CIOs and managers. Winograd and Flores claim to adhere to Heidegger's teaching, but blatantly ignore his lesson. They distinguish themselves along with all programmers and computer scientists by coining a new label: *designers*.

Heidegger's reflections regarding the deep confusion of people who are alone and lost in the wasteland that they inhabit are welcomed by Winograd and Flores as something that applies to all humans, not just them and people like them. As *designers*, they do not consider themselves to be like other humans; they live in another world, a *Sky*, from which they can observe, from high up, the people of the world they seek to redesign.[36]

As Roman historian Livy (Tito Livio) wrote, two types of tools exist: 'human tools and divine tools.'[37]

In the meta-world, Designers, super-humans or gods, and those who work within the Empire, use, at their discretion, *divina utensilia*, design instruments and codes, through which they construct *humana utensilia*, the objects and *platforms* conceded to humans, who are reduced to the role of users.

Keeping this notion in mind, Norman changes the title of *The Psychology of Everyday Things* to *The Design of Everyday Things*.

The work of Winograd and Flores, as well as Norman, has led to the creation of a plethora of segmented professional figures, each with the goal of designing one of the following: Human Interface (HI), Human-Computer Interaction (HCI),

---

[35]Terry Winograd and Fernando Flores, *Understanding Computers and Cognition*, cit., 1986, part III: Design, pp. 143–179.
[36]Francesco Varanini, *Macchine per pensare*, cit., 2016, p. 156, p. 206 and following.
[37]Livy (Tito Livio), *History of Rome (Ab urbe condĭta)*, XXXIII, in *History of Rome, Volume IX, Books 31–34* (Loeb Classical Library No. 295), Harvard University Press, 1935.

Usability, User Interface (UI), Digital Environment Designer, User Experience, (UX). A new General Theory has emerged from their work: *Design Thinking*.

This theory is the origin of a new approach to designing and planning, an approach known as Design Thinking.

The Designer, like a god, is magnanimous, and inclined to do good. But it is always easier to pretend to know more than everyone. Design Thinking considers Participatory Design to be damaging; in other words. Design Thinking is against the notion of involving all users in the design process. This theory assumes that people do not actually know what they want or what is best for them. The lives of people, their experiences, and their behaviors are nothing but "sociomaterial things" with which the Designer works.[38]

The main limitation of Design Thinking is in this view of Designers as external beings, making them Designers of the world, but not inhabitants of it. Each CIO must reflect on how easy and dangerous it can be to fall into this way of thinking.

It is important to highlight that while an architect who works in the world will necessarily clash with nature, and the world's physical constraints, a *digital worlds* Designer can construct a world from zero and plan each layer, from the foundations to the outer interface. From here we can notice a paradoxical effect: the better the structure and the platform, the more humans will feel separated from it, as humans bear the effects of new attitudes and requirements. The designer never pretends to know the people for whom he designs.

It is in this way, through Design Thinking, that this new terrain is constructed by Zuckerberg and all technicians who are dedicated to creating a digital environment in which people, as users, must live.

As Benjamin Bratton writes, through Design Thinking, we have gone "From User-Centered Design to the Design of the User.[39]" Personally, I do not believe that CIOs who work using the humanistic approach want this, however; they do not want to reduce the people who live in the digital world to the position of a user whose behavior is limited to what is considered appropriate by the Designer.

8 Human, All Too Human

If we concede to being humans, without alibi, without Divine Escape, without becoming a detached Designer of the world, we can fully appreciate Heidegger's lesson: *Thrownness* is the situation in which each human being finds himself. Each person is destined to uncertainty, annihilated by the absurdity of the world in which he lives. Each person must live in an unknown world, a wasteland. This is true of every person in every place, in every moment of human history. But this aspect of the *human condition* is even more striking during the digital disruption, when

---

[38] A. Telier (Thomas Binder, Giorgio De Michelis, Pelle Ehn, Giulio Jacucci, Per Linde, Ina Wagner), *Design Things*, The MIT Press, Cambridge, Ma., 2011, p. 6.
[39] Benjamin H. Bratton, *The Stack*, p. 284.

people live in the Stack, on the platform, experiencing a world and a way of life that is new and unknown.

The French philosopher Michel Serres, member of the Académie Française and instructor at Stanford University, marvels at the habits of his young students who belong to the generation we often call Millennials, Post-Millennials or digital natives. He is astounded by their ability to rapidly write messages using their thumbs on tiny keyboards and also by their ability to create a collective identity, through Facebook or similar platforms. The skillfulness that Serres sees in these digital natives is actually rather superficial. Adroitly moving one's fingers does not indicate that one knows what he is doing. As such, it makes sense to suspect that digital natives lack technical knowledge as well as mindfulness. Having lived through the birth of the digital world, they know how to quickly write on touch screens, communicate using WhatsApp, and express themselves on Facebook[40]. Still, knowing how to use Apps signifies very little. It is quite probable that Serres' marveling is just projection, the result of his own capacity for understanding the nature of digital culture. Serres hopes, in vain, that young people at least understand what the digital culture is all about.

Still, we cannot blame Millennials, nor the philosopher. The fact is that, when it comes to the digital world, we are all strangers, newcomers, and outsiders; therefore, with few exceptions, all of us can only act in the limits of what is defined as acceptable for users. Far from being powerful actors, we are passive beings: recipients of care, inhabitants of a platform created by a Designer. The attention given by the Designer to the user does not qualify as care. Heidegger teaches us that *care* can never be separated from beings-in-the-world.

Furthermore, digital natives, the young, the old, students, philosophers, these are not categories that refer to mindful human beings, these categories refer to the masses. Heidegger, in *Everyday Being-one's-Self*, suggests that we distance ourselves from the notion of being indistinguishable from the "große Haufen.[41]" We can translate this simply as *great masses*, but, like always, Heidegger chose his words intentionally. The word *große* reinforces the *Haufen*, which we can translate as 'accumulation', 'cluster', 'pile.' Being part of a mass of users' results in us losing our identity. The *Haufen* harkens back to the Stack as well, as a structure made from intermixed layers. Bratton points out that, as users, we make up an important part of the Stack. If we accept this condition, we agree to be part of the indistinct digital matter and it will be impossible for us to construct and use digital tools that are sufficient for meeting our needs and desires.

The CIO is particularly concerned by this *everyday Being-one's-Self*. For him, this implies constantly *Being-one's-Self* in every moment of his life, and never forgetting his professional role nor separating his work life from the rest of his life.

---

[40]Michel Serres, *Petite Poucette*, Editions Le Pommier, Paris, 2012. English translation: *Thumbelina: the culture and technology of millennials*, Rowman & Littlefield International, New York, 2015.

[41]Martin Heidegger, Sein und Zeit, § 27. English edition, cit. p. 164.

Only by agreeing to accept the difficulties that come with living a digital life on a daily basis can one distance oneself from the artificial limits placed by the Designer. By doing so, the CIO can avoid ending up just a simple regulating agent within the Stack that could eventually be replaced by a software CIO or algorithm CIO.

In the wake of Heiddegger's work, the French philosopher Derrida speaks about Thrownness, the sensation of being thrown into an unknown world by way of a narration, like with Robinson Crusoe.[42] Being shipwrecked on an unknown island is a good representation of the Thrownness felt by being stuck living in the digital world. We need to recall the story of Robinson Crusoe and put ourselves in his shoes.[43]

Robinson's loneliness, the result of being abandoned on an unknown island, manifests itself in the lack of tools. The tools that he managed to save from the sunken ship are from another world and not adapted to his situation.

Robinson is not a user. He cannot allow himself to be one, as he has nothing at his disposal that would be useful in the new world in which he finds himself. The situation forces him to start over and re-experience the Readiness-to-hand. He only has hammers, cords and some objects from another world. The first indication of his ability is how skilled he is in the domain of ready-to-hand. Robinson's ability lies in using his hands and the other-world tools to create tools that are better adapted to the new world.

Here we can look back to the idea proposed by McLuhan and Leroi-Gourhan: the tool is "an extension of man," it is a medium for establishing a relationship with the world and the environment. Only by being in a world can we create tools adapted to that world. Only by feeling at home in the new world can be appreciate what the new world offers.

Digital is unexpressed potential that can manifest itself if the appropriate instruments are in the hands of the people. We are all learners when it comes to digital tools. These tools can be constructed and used with the active participation of the digital citizen when he is no longer just a user.

We can distinguish between the Designer and the Programmer. The Programmer acts according to an analysis, looks at an objective, understands the code, and knows how to write good code. The Designers acts according to a vision, follows intuition, addresses weak points, and uses and mixes various codes.

To better understand the difference between the Designer and the Programmer, we can consider the Designer as a bricoleur.[44] This French word, *bricoleur*, harkens back to an idea of "going in a roundabout way." It has to do with knowing how to get by, even though improvising, using tools in an unexpected way. The best

---

[42]Daniel Defoe, *The Life and Strange Adventures of Robinson Crusoe, of Tork, Mariner*, London, 1719.

[43]Jaques Derrida, *Séminaire. La bête et le souverain*. Vol. II (2002-2003), Editions Galilée, Paris, 2010. English Editon: *The Beast and the Sovereign*, Volume II (The Seminars of Jacques Derrida), translated by Geoffrey Bennington, University Of Chicago Press, Chicago, 2011.

[44]Claudio Ciborra, "From thinking to tinkering". In Claudio Ciborra and Tawfik Jelassi (eds), *Strategic Information Systems*, John Wiley & Sons, Chichester, 1994.

English translation for the French word *bricolage* is probably *tinkering*. A *tinkerer* works in an amateurish or desultory way, adjusting or mending machinery, discovering solutions while working. At first glance, Robinson appears to be a *bricoleur*. He works using what he has at his disposal. He adapts tools based on his needs. But Robinson is not a Designer. A Designer works in the Sky reserved for Designers, an area closed off to users. The Designer works with tools—*divina utensilia*—that users would never have access to. Robinson, however, is on the ground, not in the Sky, and not alone: he is at a far end of the world, in a waste land. Robinson does not use *divina utensilia*, but rather *humana utensilia*, tools that are available to all people.

Even the CIO is a bricoleur. His role puts him on a different level from the Programmer. The CIO works according to his own strategy, and uses the appropriate codes and tools. The CIO uses the tools and codes at his disposal, sometimes in unexpected ways.

But being a bricoleur does not suffice. The CIO who wants to fully integrate into the new digital world will have to come down from the Designer's Sky and agree to be shipwrecked, like Robinson, on an unknown island. The CIO must concede to being a Designer and a user at the same time, and end up feeling mostly like a user.

The CIO likely has only one effective strategy for ensuring a new human environment to the people in the digital world. Above all, the strategy consists of feeling like one of the people and living their experiences firsthand; he must try to understand their resistances and put himself in the place of a user. Only if the CIO experiences what it is like to live, communicate, and work in the Stack, in the Infrastructure, in a world of Google, Facebook, and Amazon, he is able to accompany other people in this new world.

9 A Rebirth of Literacy

We must not forget Bratton's political literature regarding the digital revolution: we must look at the Stack as reducing personal liberties and subordinating people to an entity that is both oppressive and invisible, as is typical in the context of globalization and financial deregulation.

From this perspective, the digital revolution is something of a manifestation of *The End of History* and *Postmodernity*. The human being seems unable to control new technology. The future and progress seem lost in an eternal present. Authority, power, and sovereignty reside in the infrastructure—those who govern the infrastructure govern the world. Here we see the political importance of the CIO.

But this phenomenon must not cause us to forget that the digital revolution exists as a humanistic project, rooted in a historical period with faith in progress and a new type of humanism: the 1960s, a period of growth and change. The decade begins with a departure of the gloomy atmosphere brought about by the Cold War. Kennedy's New Frontier opens new horizons. A new hope for peaceful coexistence took the place of a fear of nuclear conflict. The Vietnam War led to a new pacifism.

Widespread prosperity lead to greater attention to immaterial needs and an increase in personal liberties. "The message," according to poet Allen Ginsberg, was "to widen the area of consciousness.[45]" These were the years of Counterculture, the Youth Revolution, the Movement, and Sex, Drugs, Rock and Roll.

Many things happened in this decade. In the field of information technology, advances were made in Artificial Intelligence. Meanwhile, IBM introduced the Mainframe System/360 to the market, the first group of computers capable of responding to the needs of a business, furnished with interchangeable software and peripheral equipment. It was a digital machine, but it had nothing to do with what we now call the Digital Revolution.

The Digital Revolution is strongly connected to the ideas regarding technique proposed by McLuhan, Leroi-Gourhan and even Heidegger. This 'humanistic' technique, as we have seen, offers tools that could be an extension of man—of which the Mainframe is the most obvious example. The cost of the various technical components, meanwhile, was constantly decreasing. Widely distributed journals, such as Popular Electronics, talked about computer architects as hobbyists, bricoleurs, and tinkers, capable of building their own computers. As a result, in this period, visionary technicians built the prototypes for the tools that we now know as the network of Personal Computers, which includes the desktop, laptops, tablets, and smartphones—tools that are increasingly ready-to-hand.

Joseph Carl Robnett Licklider, known as J.C.R. or Lick, was a psychologist specialized in psycho-acoustics. In the 1950s, he developed an interest in computers while working for the Lincoln Laboratory at the Massachusetts Institute of Technology, a research center funded by the Department of Defense.

In 1960, he published a short article: *Man-Computer Symbiosis*.[46] The reference to symbiosis is particularly interesting: Licklider explains that it is a biological concept. In his description, Licklider defines two different classes of organisms, "human brains and computing machines," that can live "together in intimate association, or even in close union." "The hope is that, in not too many years, will be coupled together very tightly." "The resulting partnership will think as no human brain has ever thought and process data in a way not approached by the information-handling machines we know today."

Given these assumptions, Licklider continues, "it seems reasonable to envision, for a time 10 or 15 years hence, a 'thinking center' that will incorporate the functions of present-day libraries together with anticipated advances in information storage and retrieval." Licklider does not stop there: he has a clear idea in his head of an infrastructure. "The picture readily enlarges itself into a network of such centers, connected to one another by wide-band communication lines and to individual users by leased-wire services."

---

[45]Allen Ginsberg, *Kaddish and Other Poems 1958–1960*, Pocket Series 14, City Lights, San Francisco 1961, Note, p. 100.
[46]J. C. R. Licklider, "Man-Computer Symbiosis", in: IRE (Institute of Radio Engineers) *Transactions on Human Factors in Electronics*, volume HFE-1, pp. 4-11, March 1960.

Licklider is able to do more than just conceive anticipatory visions; going beyond his training in psychology, he understands quite well the technical and economic aspects of these visions: "in such a system, the speed of the computers would be balanced, and the cost of the gigantic memories and the sophisticated programs would be divided by the number of users."

His management capacities were also recognized early on. In 1962, Licklider was nominated as the head of the Information Processing Techniques Office (IPTO) of the Advanced Research Projects Agency (ARPA); like NASA, the ARPA was created in 1958 in response to Soviet Union launching Sputnik into space. In the ARPA, Licklider was in charge of the IPTO. In 1963, he became the ARPA's Director of Behavioral Sciences Command & Control Research.

A memorandum that was circulated on April 23, 1963, suggested connecting, in one network, all computers involved in research and development projects—called the *Intergalatic Computer Network*.[47] According to Licklider, only by creating such a network and sharing hardware resources and calculation power, can the "aspirations, efforts, activities" of all those involved in the various projects—"advancement of the art or technology of information processing", "advancement of intellectual capability (man, man-machine, or machine)"—be achieved and completed with success.

With these words, Licklider provided the first description of the ARPAnet, which first appeared in 1969 and would eventually be renamed Internet. Still, it is clear that Licklider did not view the Internet as just an underlying internal structure, a physical layer. Licklider already had a clear vision in mind for the social capabilities of the Internet, the layer of shared knowledge: The World Wide Web.

Licklider did not stop there. In *Libraries of the future*,[48] the final report of a research project commissioned by the Ford Foundation, he discussed an issue of great interest even today, fifty years later: the digitization of libraries, books, and paper archives, and thus, the appearance of the digital Revolution and the conservation of knowledge through digital storage media. Even in the mid—1960s, Licklider had already noticed how the amount of material that people were trying to conserve was growing exponentially. This increase was accompanied by storage media's growing capacity. Even the structure of texts was affected as, with digital supports, they could always be accompanied by metadata and annotations. Now, we will talk about Document Management, Full Text Indexing, Information Retrieval, Search Engine Technology, Multimedia Objects, and Knowledge Objects: all of these were discussed, in some detail, in *Libraries of the future*.

Licklider did not stop there. In his 1968 article, *The Computer as a Communication Device*,[49] he offers a detailed preview of what we have come to call

---

[47] J. C. R. Licklider, "Memorandum For Members and Affiliates of the Intergalactic Computer Network", April 23, 1963. Published on *KurzweilAI.net*, December 11, 2001, http://www.kurzweilai.net/memorandum-for-members-and-affiliates-of-the-intergalactic-computer-network.

[48] J. C. R. Licklider, *Libraries of the future*, The MIT Press Cambridge, Ma., 1965.

[49] J. C. R. Licklider, "The Computer as a Communication Device", *Science and Technology*, April 1968.

the *web 2.0 infrastructure, social network, virtual community*. Licklider called these "on-line interactive communities." These communities contained people who wished to collaborate "face to face, through a computer."

"What will on-line interactive communities be like?" Licklider wondered. "In most fields, they will consist of geographically separated members, sometimes grouped in small clusters and sometimes working individually. They will be communities not of common location, but of common interest." Licklider also points out that "in each field, the overall community of interest will be large enough to support a comprehensive system of field-oriented programs and data."

We must still hope that this concept will come to fruition because, in the mid—1980s, Winograd and Flores affirm the dominion of the Designer, who constructs programs and data from the Sky and not from within the interactive communities. It is for this reason that even 50 years later, interactive communities are largely constricted to Facebook and other similar contexts and subject to a system of rules.

For Licklider, the threats and opportunities are already quite clear. The threat comes from a new Social inequality: Digital divides. Hope appears in the form of a Digital Democracy, a New Economy founded on Knowledge Sharing.

For the society, the impact will be good or bad, depending mainly on the question: Will "to be on line" be a privilege or a right? If only a favored segment of the population gets a chance to enjoy the advantage of 'intelligence amplification', the network may exaggerate the discontinuity in the spectrum of intellectual opportunity.

On the other hand, if the network idea should prove to do for education what a few have envisioned in hope, if not in concrete detailed plan, and if all minds should prove to be responsive, surely the boon to humankind would be beyond measure.

Licklider's mentioning of 'intelligence amplification' is a direct reference to the work of Doug Engelbart. Licklider, in his role as IPTO of the ARPA financed—along with NASA and RADC (Rome Air Development Center's Research and Development Laboratory, part of the U.S. Air Force)—The Augmentation Research Center (ARC) of the Stanford Research Institute, Menlo Park, California, which was Engelbart's laboratory. Engelbart, an electrical engineer, is also a technician, machine builder and implementer. He is also a visionary. He thought of Augmenting Human Intellect through human interaction with machines.

The 1962 request for funding was the chance to describe the *Conceptual Framework* of the work.[50] At work, and even in one's personal life, everyone has to deal with "complex situations," without clear or linear solutions. "By 'augmenting human intellect,'" wrote Engelbart, "we mean increasing the capability of a man to approach a complex problem situation, to gain comprehension to suit his particular needs, and to derive solutions to problems". For Engelbart, increasing one's

---

[50]Douglas Engelbart, *Augmenting Human Intellect: A Conceptual Framework*, Summary Report Prepared for Direction of Information Science Air Force Office of Scientific Research, Stanford Research Institute, October 1962.

capabilities means a mixture of "more-rapid comprehension, better comprehension, the possibility of gaining a useful degree of comprehension in a situation that previously was too complex, speedier solutions, better solutions, and the possibility of finding solutions to problems that before seemed insoluble."

Engelbart points out that, faced with all of these complex problems, one needs more than "isolated clever tricks that help in particular situations." One must completely change his "way of life." People need to prepare to live in an "integrated domain," where "hunches, cut-and-try, intangibles, and the human 'feel for a situation' usefully co-exist with powerful concepts, streamlined terminology and notation, sophisticated methods, and high-powered electronic aids." This is the digital world: a world where human beings and computers, connected by "man-artifact interfaces," make up one system.[51]

In order for the system to be effective, the machines cannot resemble the Mainframes or the Infrastructure, both far separated from people; the machine must be a ready-to-hand tool that is easily manipulated by people and that can help people face complex problems.

At this point, we must ask ourselves how we can imagine and build ready-to-hand tools. In the mid-1980s, we saw how Winograd and Flores, perhaps guided by the best intentions, described a world with a Designer, and everyone else existing only as users. In doing so, they created a world in which the Designer's tools were necessarily different from the users' tools.

In October 1962, Engelbart wrote about a very different approach:

(1) Our researchers are developing means to increase the effectiveness of humans dealing with complex intellectual problems, and (2) our researchers are dealing with complex intellectual problems. In other words they are developing better tools for class to which they themselves belong.[52]

Engelbart's Research does not live in a separate world, like the Designer's Sky proposed by Winograd and Flores; the Research, instead, lives in the *throws*—as Heidegger tells us—in the same world, the same environment as the human being. The Researcher, as a human, can work effectively only if he always remembers that the tools he uses can also be used by any other human being. Each human being who performs the job of Researcher, or any other job, faces complex problems: he can therefore take advantage of digital tools.

Thus, two different attitudes exist. Each CIO must decide which attitude to assume, either the Designer of Winograd and Flores, or the Researcher of Engelbart.

December 9, 1968, San Francisco, Monday Afternoon, 3:45.

The afternoon session of the Fall Joint Computer Conference begins, held at the Convention Center in San Francisco. More than 1000 computer professionals are in attendance.[53]

---

[51]Douglas Engelbart, *Augmenting Human Intellect*, cit., p. 1.
[52]Douglas Engelbart, *Augmenting Human Intellect*, cit., p. 118.
[53]Available at: http://www.youtube.com/watch?v=yJDv-zdhzMY.

Six years have passed since Engelbart wrote *Conceptual Framework*, six years of technical work in a laboratory. The conceptual framework has not changed and the visionary hope persists, but now everything has been brought to life by practical experience. Not the pure and abstract experience of the researcher, but a type of completely human experience, connected to man's daily life, for each man in the world who has chosen not to submit to the will and orders of others.

Instead of studying how tools could enhance the operations of the human mind, years have been spent just *using tools*. In their use, tools refine and the mind finds ways to work better. The practice feeds the theory and theory, in turn, feeds into the practice.

Smiling, though not without worry, Doug Engelbart starts to talk.[54] He had a strange machine in front of him, "a computer-based, interactive multi-console display system", a machine that is completely at man's disposal, his mind's prosthesis, a machine connected to other machines, a knot in an infinite network. Today, we call it the Personal Computer.

Engelbert showed how the Word processor worked. This way of writing that is commonplace now, was so different from writing on paper. "Word processing beginning with a blank piece of paper," Engelbart says. Engelbart writes on the screen and explains "An instrument/vehicle for helping humans to operate whitin the doman of Complex Information Structure". *Operate*, he tells us, is "compose, study and modify." The Complex Information Structure is—he shows us while tracing a graph on the screen—a representation of links between concepts. The structure that we have in mind "is too complex to investigate in linear text," and so we need more than a machine that just processes texts made of sequenced words. Engelbart uses the mouse, and talks about it. The first mouse ever seen. "I don't know why we call it a mouse," he says. "It started that way and we never changed it." He and his collaborators invented the tools and gave them names.

Engelbart connects his machine with his collaborators' machines, who are at the Menlo Park laboratory: we see their images on the screen, on Engelbart's shoulders while he speaks with them. The infrastructure of Engelbart, named NLS, oN-Line System, couples two types of work: off-line workflow and online cooperation.

The description of what is now called the Personal Computer was already clear in the Computer Display Control Report, given to the investors—ARPA, NASA, and RADC—in 1965: a workstation with a cathode ray tube screen, keyboard, pushbuttons, mouse, and a joystick. Human beings interact with the machine through convenient interfaces. "A user soon finds it very easy to keep his eyes on the screen and cause the bug to move about upon it as quickly and naturally as if he were pointing his finger (but with less fatigue).[55]" Tools ready-to-hand. Someone mentions the possibility of failure, as the tools developed by Engelbart's Augmentation Research Center never resulted in the creation of a business. For

---

[54]The Mother of All Demos, https://www.youtube.com/watch?v=yJDv-zdhzMY

[55]W. K. English, D. C. Engelbart, *Computer-Aided Display Control*, Stanford Research Institute, Menlo Park, July 1965, p. 6.

various reasons, the tools were never patented, or else the patents were never defended. The Augmentation Research Center, due to a loss in funding, entered a period of crisis. Taking advantage of the ARC's difficulties, Xerox was able to open a new research center in Palo Alto, a few miles from Menlo Park, in 1970. The Palo Alto Research Center, PARC, is considered the cradle of digital technology, but only because they received many of the researchers who originally worked for Engelbart.

The vision of "the digital computer as a tool for the personal use of an individual," offered by Engelbart is strong, because it comes from a solid thought process. Engelbart, an electrical engineer, was actually a philosopher, and, like Leroi-Gorhan and McLuhan, was capable of understanding how technique led to the "natural evolution in developing the basic human capabilities." Therefore, "in a very real sense, as represented by the steady evolution of our augmentation means, the development of artificial intelligence has been going on for centuries." That which Engelbart called "augmentation mean" was what McLuhan called "media, extension of man." A trick on nature carried out by man.

Engelbart, writing in the late 1960s, reminds us that the word processor and the mouse are augmentation means. But he does not limit himself to philosophy, he builds tools.

The CIO is not asked to build tools. He is, however, asked to choose tools. He must choose them in a way that gives all people the possibility to expand their intelligence and mindfulness.

Ted Nelson was born in 1937. His parents were film actors in their 20s. His father became well-known as a director. His mother, actress Celeste Holm, became even more famous and won an Oscar. They did not have much time for their son. Ted grew up with his maternal grandparents: his grandmother was an eccentric artist.

Ted spoke without hesitation of the "sickness" that accompanied him in his youth and, in different ways, into adulthood. Little Ted was dyslexic, unable to read printed texts at the speed that his teachers expected. It ended up that the "sickness" was more than just dyslexia: Ted suffered from Attention Deficit Disorder. In medical and psychiatric jargon, two conditions are universally known, yet subtlety different: ADD: Attention Deficit Disorder; ADHD: Attention Deficit Hyperactivity Disorder.

"Attention Deficit Disorder," explained an older interviewer, "was coined by regularity chauvinists. Regularity chauvinists are people who insist that you have got to do the same thing every time, every day, which drives some of us nuts. Attention Deficit Disorder: we need a more positive term for that. Hummingbird mind, I should think.[56]"

From early childhood, for Ted, the traits typical of written text posed insurmountable difficulties. They impeded his learning. A sheet of paper was a weak support that lacked a third dimension. Indeed, writing has just one dimension: sequenced characters on a support, in rigid order, letter after letter, word after word,

---

[56] Gary Wolf, The Curse of Xanadu, Wired, 3.06, June 1995.

sentence after sentence, line after line. Sheets of paper are bound in notebooks or books, and books are often found, in some order, in libraries.

Teachers and doctors consider dyslexia and ADD to be defects, sicknesses. The Hummingbird mind needs to be corrected and brought back to a normal rhythm. But Nelson thought otherwise: "It was just another way of cognitive processing. A different mode of thinking."

Young Nelson, dreamed of a way that learning that was less harmful. He dreamed of a type of technology that, instead of being critical of his deficiencies, would help him and value his individuality. What I feel so strongly, Nelson thought, must also be felt by others. Anyone could benefit from a new way of learning that did not force learners to rigidly follow a set structure, a way of learning that was not limited to closed blocks, book pages, and libraries.

He dreamed of a type of fluid knowledge that created new pathways, in which each place written about by people throughout history was connected to all other places.

We must also point out that it was not only sequential writing, pages, books, and libraries that Ted struggled with. He also struggled with the order imposed on all Information Systems. Records management, Data modeling, Data Base Management Systems are all attempts at putting data in order. These were just a new twist on the sequential ordering systems that were already noticeable in writing, pages, books, and libraries. Nelson was not trying to say that all of this was useless. But he did feel that alternatives existed. The data in all Information Systems are not on paper, but rather on digital supports. But Nelson points out that this transition was not the real cultural shift. The shift was a change in how people read and wrote: writing on a word processor liberates us from having to write sign after sign, liberates us from cutting and pasting. The shift was in imagining the fruit of all human writing as one lone text-here, we think of the word 'seamless.' A text that was not on paper, but rather, was multidimensional. Today, we often use the word *hypertext*. It is a new word invented by Ted Nelson in the early 1960s.

In his 20s, Ted Nelson, having extensively studied literature, comes face-to-face with his childhood need for inclusion while reading the poetry of Coleridge. Coleridge presents a dream—"A vision in a dream"—Xanadu, a palace on a river in the Orient, a magnificent world.[57] Ted Nelson imagines his own Xanadu. He dreamt of a different way of "reading." A way that did not involve printing. A way that did not even go through Structured Information Systems.

Starting in the early 1960s, Nelson would spend his whole life trying to realize his dream. A form of technology that would allow for a new type of literature, in which all texts would be connected to all other texts, in full transparency; a type of literature in which the contributions of all authors are weaved together in a big, collective text.

---

[57]Samuel Taylor Coleridge, *Kubla Khan. Or a Vision In a Dream. A Fragment* (1816), in Samuel Taylor Coleridge, *Christabel, Kubla Khan, and the Pains of Sleep*, John Murray, London, 1816, pp. 51–58.

Some say that Nelsons projects did not end in failure. Even today, Nelson, now much older, continues to advance his projects, and is still facing new challenges. He considers the World Wide Web a "draft" of his project. But the World Wide Web would have never existed if Nelson did not have the vision of a new culture, a new digital culture. Tim Bernes Lee, who we know as the inventor of the World Wide Web, refers to Ted Nelson as his first role model.

Similarly, several others refer to Ted Nelson as their role model, people such as Bill Gates, Steve Jobs, and all the young, enthusiastic Programmers, those who became adults in the 1960s, to whom we credit with creating the tools that started the Digital Revolution. All were inspired by a book published by Ted Nelson in 1974: *Computer Lib/Dream Machine*.[58] A book that is the coming-of-age story of this generation of innovators. A book without pagination, written by hand and illustrated by the author, made up of a collection of typed texts glued together. It was an Underground text, an artists' book similar in form to the work of cartoonist Robert Crumb.

The book's cover alone was provocative and suggestive: a human hand, closed in a fist, with the word Lib on the write and, higher, the word Computer. At the top, the following sentence was written: "Can and must understand computers. NOW."

EVERYBODY SHOULD UNDERSTAND COMPUTERS, Nelson wrote in the introduction, in all capital letters. This book "is intended to fill a crying need. Lots of everyday people have asked me where they can learn about computers, and I have had to say nowhere."

"I was a very junior computer programmer and occasional teacher of Transcendental Meditation," recalls Mitch Kapor, Designer of Lotus 1-2-3. "I stumbled upon Computer Lib on a nocturnal excursion and was instantly bewitched. Here was a man who dreamed my dreams before I did, who gave voice to a radically different concept of computers as other than giant calculating machines."

Ten years later, in the early 1908s, in Literary Machines, Nelson spoke of two hopes:

Hope 1. To have our everyday lives made simple and flexible by the computer as a personal information tool.
Hope 2. To be able to read, on computer screens, from vast libraries easily, the things we choose being clearly and instantly available to us, in a great interconnected web of writings and ideas.[59]

"The immensity of the coming revolution is not clear yet," Nelson writes. Perhaps the immensity of the Digital Revolution is still not clear today, thirty years later.

---

[58]Ted H. Nelson, Computer *Lib/Dream Machines*, Self-published, 1974. Second Edition: Tempus Books/Microsoft Press, Redmond, Washington, 1987.
[59]Theodor Holm Nelson, *Literary Machines*, Self-published), Swarthmore (Pa) 1981, 1/2. (Editions as listed in the 93.1 (1993) edition: 1980, 1981, 1982, 1983, 1984, 1987, 1990, 1991, 1992, 1993).

According to Nelson, this immensity remains unclear because "two cultures have united on a false, agreed-upon definition of what computers are."

On one side are the technicians, who Nelson calls *Technoids*, or, *Noids*. They "have an exaggerated and caricatured notion of what constitutes clear-minded thinking, and never miss a chance to denounce other cognitive styles al 'illogical.'" Nelson acknowledges the irony: "Technoids are Lords of Complication."

Still, Nelson is equally critical about members of the other culture: the *Fluffies*.

Fluffies have "a humanistic background, in literature, history, the arts, etc." "The Fluffy cognitive style leans toward vagueness and the reduction of issues to vague idealistic terms."

"The members of the two cultures, technical and literary—who rarely talk to each other—", concludes Nelson, "have it all figured out, quite wrongly.[60]" A new culture is needed, one that can find a middle ground between these two.

"The goal of tomorrow's text systems will be the long ones of civilization—education, understanding, human happiness, the preservation of human tradition." "But", adds Nelson "we must use today's and tomorrow's technologies."

A new profession, a new role is needed. Nelson offers a name for this role, *System Humanist*. The System Humanist, writes Nelson, must strive for "the ideals of the humanist perspective by the best available means. This means finding the way the human literature, art and thought—including science, of course—may best be facilitated, preserved, and disseminated.[61]" "Means to increase the effectiveness of humans dealing with complex intellectual problems," Engelbart wrote. Nelson and Engelbart knew each other well; already in the 1960s, they shared the same vision. Means, or media, or "extension of man." Humanistic, digital tools.

A CIO can and must be a System Humanist: far from the attitudes of the Lords of Complication, attentive to education, understanding, human happiness, the preservation of human tradition, and ready to use today's and tomorrow's technologies.

10 The Human Being Does not Work that Way

Licklider, Engelbart, and Nelson, our trailblazers—pioneers, guides on the path to digital mindfulness—claim to closely follow a trailblazer, role model and master, who belong to the previous generation: Vannevar Bush.

It all started with an article entitled *As We May Think*, published in the journal *The Atlantic* on July 25, 1945. A second version, shortened but with illustrations, appeared two months later, on September 10, in the weekly publication, *Life*.

Engelbart read *As We May Think* in *Life*, in September 1945 while, during the war, serving in the Philippines, as a radar technician for the United States Navy.

---

[60]Ted Nelson, Literary Machines, cit., 1/11.
[61]Ted Nelson, Literary Machines, cit., 1/13.

The article captivated him. The article would later be the main inspiration of one of Engelbart's projects: *Augmenting Human Intellect*. Licklider dedicated *Libraries of the Future* to Bush. Nelson showed his respect and gratitude with "an effort in counter-discipleship," an article with an emblematic title: *As we will think*. Here, at the beginning of the 1970s, Nelson argued: Bush's work is a necessary starting point for making our vision for the machine a reality, "that much of what he predicted is possible now." Nelson later republishes all of *As We May Think* in *Literary Machines*.

Bush was an avid supporter of technical innovation: science and industry together, basic research but also with tendencies towards the killer app and practical use. When he was young, Bush worked for General Electric. Then he graduated from MIT with a degree in Electrical Engineering. He became a researcher there in 1919, then a professor in 1923, and between 1932 and 1938, he was Vice President and Dean.

During this time, he was also an entrepreneur. In 1922, he was one of the founders of Raytheon. Initially, the company produced electron tubes to supply power to radio-receivers, making it possible to convert alternating current to direct current. The electron tubes would next be used during World War II in the construction of radar. Today, the Raytheon Company is a major U.S. defense contractor.

Starting in 1927, Bush directed the MIT laboratory in which the *differential analyzer* was constructed, designed to solve differential equations by integration. Bush was quite familiar with punch card technology, through which information could be digitized: represented by a series of numbers. Bush was also quite familiar with electron tubes. Still, the differential analyzer was neither digital nor electronic. It was a mechanical engine, using wheel-and-disc mechanisms.

In the 1930s, experiments were conducted on the use of electronic components: electron tubes, relay contacts, and switches. Through these experiments on analytical engines, Claude Shannon, a twenty-year-old student, conceived an essential core of the digital computer. It was a structure composed of two layers. The Logic layer: information was expressed as a numeric code to the second base: a binary digit: a bit. The Physical Layer: each bit corresponded to an electronic circuit, either an open or closed state. This description was the main idea of the thesis for Shannon's Master of Science.

In the second half of the 1930s, in a time of global crisis and impending war, Bush fostered a government agency aimed at coordinating and directing the technological innovations of universities and private businesses, which could be used for military purposes.

In an early 1940 Congressional meeting, the discussion regarding the National Defense Research Committee (NDRC) was proceeding slowly. In May, when Germany invaded France, Bush used lobbies and private contacts to organize a meeting with Roosevelt. On June 12, 1940, the president approved and supported Bush's vision.

In 1941, the NDRC was absorbed by the Office of Scientific Research and Development (OSRD).

Bush, as director of the OSRD, coordinated more than 200 scientific, military projects. Radar, sonar, computer-based spotting devices, mass production of penicillin and sulfonamides. Until 1943, when the ORSD was taken over by the army, Bush also led the Manhattan Project: the atomic bomb.

On April 12, 1945, Roosevelt died. Harry Truman took his place, though he knew little about the work on atomic bombs and was relatively unknown by Bush. On August 6, the B-29 Super fortress dropped the Little Boy on Hiroshima.

The short, informative article *As We May Think*, was published in *The Atlantic* on July 25, 1945: a few days before the bombing of Hiroshima. The second version appeared in *Life* one month after the bombing.

Bush was a master of Public Relations and Cultural Politics. The article had a precise intent. Bush, who was often at odds with the Harry Truman, was trying to convince the president of the need to focus on new research projects that could be useful in peacetime, rather than focusing only on projects for the Military-Industrial Complex. Bush wrote extensively, always avoiding technical terms, so as to connect with the masses. He considered Information Technology central to post-war society. His worked focused on how computers could assist research projects. We could easily suppose that Bush was thinking about mediating platforms that could share advances and knowledge. But, on the contrary,—perhaps as a reaction to the large dimension of military/industrial research projects, in which the researcher, deprived of a vision of the whole project, risked being reduced to the position of 'skilled worker'—Bush imagined a desk-machine, a personal machine, for individual use, for the purpose of asking questions and looking for answers.

Bush and Licklider both pointed out that we will become increasingly overwhelmed by a large mass of information. In 1945, Bush already saw the solution offered by Information Technology to the problem posed by this mass of information; it was in the same vein as what was proposed by Descartes five hundred years prior: look for an ordering system, a type of classification. Information can be broken down into data, data can then be selected and conserved in predefined spaces, then stabilized with a map or model: essentially, the Data Base Management System.

"Our ineptitude in getting at the record is largely caused by the artificiality of systems of indexing. When data of any sort are placed in storage, they are filed alphabetically or numerically, and information is found (when it is) by tracing it down from subclass to subclass. It can be in only one place, unless duplicates are used; one has to have rules as to which path will locate it, and the rules are cumbersome. Having found one item, moreover, one has to emerge from the system and re-enter on a new path.[62]"

Bush raised one simple, humanistic exception: "The human mind does not work that way." This brings us to a crossroads. On one hand, it would be possible to institute machines that guarantee order and impose rules. On the other hand, it

---

[62]Vannevar Bush, *As We May Think*, cit., 1945, § 6.

would also be possible to construct machines that more closely mirror a human way of thinking.

"The human mind [...] operates by association. With one item in its grasp, it snaps instantly to the next that is suggested by the association of thoughts, in accordance with some intricate web of trails carried by the cells of the brain. [...] The speed of action, the intricacy of trails, the detail of mental pictures, is awe-inspiring beyond all else in nature.[63]"

With this remark, Bush forces us to face the principal issue with digital culture. The Digital Revolution forces people to live in a world characterized by an "enormous mass" of data, a world made up of data, information, and bits.

Here is a new divergence: "Being digital" can mean two opposing things. On one hand, it means belonging to an Infrastructure that reduces us to the role of user, limited to certain behaviors, defined by the Designer. On the other hand, it means being thrown into the world alone to face different problems, but equipped with a machine that can help us in our personal journey: from question to answer; from a mass of information to the discovery of new ways of solving problems; from ignorance to knowledge.

An open question. A question posed by all CIOs. Bush, Licklider, Engelbart, and Nelson suggest a humanistic stance. Bush, engineer, machine builder, elects to view machines not as a builder, but from the perspective of the person who needs a tool for personal empowerment.

As a result, during the wartime of the 1940s, Bush imagines a device, a group of tools for peacetime, prostheses of our body and mind. Bush called it *memex*. Now, we say *Personal Computer, tablet, smartphone*.

"Consider a future device for individual use, which is a sort of mechanized private file and library. It needs a name, and, to coin one at random, 'memex' will do. A memex is a device in which an individual stores all his books, records, and communications, and which is mechanized so that it may be consulted with exceeding speed and flexibility. It is an enlarged intimate supplement to his memory.[64]"

By moving our hands across the keyboard or a mouse across the screen, we are enlarging the capacity of our minds; it is our minds' way of moving "on the intricacy," "on the web of trails."

Bush attached a special importance to the word *trail*. For Bush, *Trail* refers to both the connections that are activated in our brains, our personal neural networks, as well as the connections—now called *links*—that connect documents, texts, and the information at our disposal—now called the World Wide Web. Bush suggests viewing the World Wide Web and our own minds as mirrors of each other.

The verb *trail* is derived from the word *trahere*: to 'draw', 'to drag', 'to haul', 'to get', 'to derive'. Starting in the fourteenth century, it also meant 'to follow the traces or scent of, as in hunting', 'to track'. Consequently, the noun *trail* means 'a

---

[63]Vannevar Bush, *As We May Think*, cit, 1945, § 6.

[64]Vannevar Bush, *As We May Think*, cit-, 1945, § 6.

mark, trace, course, or path left by a moving body', 'track or smell left by a person or animal', 'a marked or beaten path, as through woods or wildernesses. It also refers to 'an overland route': the pioneers' trail across the prairies.

In the same period that the Digital Revolution seemed imminent, that is, the final years the twentieth century, the years in which Negroponte wrote *Being Digital*, Peter Drucker affirmed that "The most important contribution management needs to make in the 20th century is [...] the increase the productivity of 'knowledge work' and 'knowledge workers'".[65] Vannevar Bush had spoken about all of this fifty years earlier and had said it with greater clarity. Bush did not speak only about the work, just as he did not speak only of the user of an Infrastructure or platform. Bush discussed the citizen. He did not speak only about productivity, he spoke about how each human being lives in a new environment: a digital environment. Bush insists that we "find delight in the task of establishing useful trails through the enormous mass of the common record."

The vision is clear: "Wholly new forms of encyclopedias will appear, ready-made with a mesh of associative trails running through them, ready to be dropped into the memex and there amplified.[66]" This means we must fully embrace this world—not just limit ourselves to using Facebook. Thus, we are all called to be trailblazers: pioneers exploring new terrain. To do so, however, we need someone who has already started this journey and can guide us.

11 Conclusions

Such is the role of the Digital Humanist CIO. Exploring the digital land. Experimenting with different ways of living this new environment.

Digital technologies offer people a new possibility: redesigning one's entire life, from daily life to work life, from homes to businesses. Of all managers, the CIO is best equipped to understand this change. He will be the one to guide businesses and organizations through the Digital Transformation.

This task implies certain technological decisions. The easiest choice, the choice that is closest to the historical role of the CIO consists in establishing infrastructures and platforms, where each person, citizen or worker, is reduced to the role of user, whose actions are limited by certain rules. Infrastructures and platforms are necessary, but for another reason: offering all citizens and workers tools for constantly enlarging their work spaces: only in this way can we fully take advantage of the implicit richness of the Digital Revolution.

The other option, however, requires the CIO to tap into his own knowledge of ethics and the human condition. The CIO, as an expert in controlling machines, is

---

[65]Peter F. Drucker, "Knowledge-Worker productivity: The Biggest Challange", in *California Management Review*, vol. 41, n. 2, Winter 1999.
[66]Vannevar Bush, As We May Think, cit-, 1945, § 8.

called to reflect on his own experiences with digital innovations and be in constant contact with the people with whom and for whom he works.

Millennials, digital natives, citizens, workers, Designers, and CIOs—we are all like Robinson Crusoe. We must learn to get by in a new world and use new tools. To do so, however, we need someone who has already started this journey and can guide us. This is the role of the humanist CIO. The best trailblazer is the person who has ventured into the forest, risked being lost, and ultimately found his way.

References

Arendt H (1988) The human condition. The University of Chicago Press, Chicago
Bratton BH (2015) The stack: on software and sovereignty. The MIT Press, Cambridge
Burgess M (2015a) In search of certainty. The science of our information infrastructure, O'Reilly, Sebastopol
Burgess M (2015b) Thinking in promises: designing systems for cooperation, O'Reilly, Sebastopol
Bush V (1945) As We May Think. The Atlantic
Cabitza F, Varanini F (2017) Going beyond the system in systems thinking: the cybork. In: Rossignoli C, Virili F, Za S (eds) Digital technology and organizational change: reshaping technology, people, and organizations towards a global society, Proceedings ITAIS 2016, Lecture Notes in Information Systems and Organisation (LNISO). Springer, London
Ciborra C (1994) From thinking to tinkering. In: Ciborra C, Jelassi T (eds) Strategic information systems. Wiley, Chichester
Coleridge ST (1816) Kubla Khan. Or a Vision In a Dream. A Fragment, in Coleridge ST, The Complete Poems. In: William Keach (ed) Penguin Books, London, 2004
Defoe D (1719) Robinson Crusoe, Wordsworth, Ware, Hertfordshire, 1997
Derrida J (2011) The beast and the sovereign, vol II (The Seminars of Jacques Derrida). University of Chicago Press, Chicago
Drucker PF (1999) Knowledge-Worker productivity: The Biggest Challange, in California Management Review, vol. 41, n. 2, Winter
Engelbart D (1962) Augmenting human intellect: a conceptual framework. Summary report prepared for direction of information science air force office of scientific research. Stanford Research Institute, Menlo Park
Engelbart D (1968) The Mother of All Demos. http://www.youtube.com/watch?v=yJDv-zdhzMY
English WK, Engelbart D (1965) Computer-aided display control. Stanford Research Institute, Menlo Park
Ginsberg A (1961) Kaddish and other poems 1958–1960. In: Pocket series 14. City Lights, San Francisco
Heidegger M (1996) Being and time. University of New York Press, Albany
Leroi-Gourhan A (1993) Gesture and Speech. MIT Press, Cambridge
Licklider JCR (1960) Man-computer symbiosis. In: IRE transactions on human factors in electronics, volume HFE-1
Licklider JCR (1965) Libraries of the future. The MIT Press, Cambridge
Licklider JCR (1968) The Computer as a Communication Device. Science and Technology
Licklider JCR (2001) Memorandum for members and affiliates of the intergalactic computer network (April 23, 1963). KurzweilAI.net, December 11, http://www.kurzweilai.net/memorandum-formembers-and-affiliates-of-the-intergalactic-computer-network
Livy (Tito Livio) (1935) History of Rome vol IX, (Loeb Classical Library No. 295). Harvard University Press, Cambridge, pp 31–34

McLuhan M (1964) Understanding media. The extension of man. MacGrow Hill, New York
Negroponte N (1995) Being Digital. Knopf, New York
Nelson TH (1987) Computer Lib/Dream machines. Microsoft Press, Redmond
Nelson TH (1993) Literary Machines. Self-published, Swarthmore
Norman D, Draper S (eds) (1986) User centered system design: new perspectives on human-computer interaction, L. Erlbaum Associates, Hillsdale
Norman D (1990) The design of everyday things. Currency Doubleday, New York
Serres M (2015) Thumbelina: the culture and technology of millennials. Rowman & Littlefield International, New York
Telier A (2011) Design things. The MIT Press, Cambridge
Varanini F (2012) Complexity in projects: a humanistic view in Francesco Varanini & Walter Ginevri, projects and complexity. CRC Press, Boca Ratón
Varanini F (2013) Il ricercatore debole, o la restituzione poetica, in Bocchi G, Varanini F, Le vie della formazione. Creatività, innovazione, complessità, Guerini e Associati, Milano
Varanini F (2016) Macchine per pensare. L'informatica come prosecuzione della filosofia con altri mezzi, Guerini e Associati, Milano
Winograd T, Flores F (1986) Understanding computers and cognition: a new foundation for design. Addison-Wesley, Reading
Wolf G (1995) The curse of xanadu. Wired, 3.06
Yates F (1966) The art of memory. Routledge & Kegan Paul
Zuckenberg M (2017) Building global community. https://www.facebook.com/notes/mark-zuckerberg/building-global-community/10103508221158471/. Accessed 16 Feb

The CIO and the Digital Challenge

Daniele Rizzo

Abstract Digital native organisations are setting new rules and expectations around Information Technology. Not only they are disrupting industries and business models, but also the IT practice itself. Their methods and standards are discontinuing older, established enterprise IT disciplines (like project management, system design, etc.). CIOs and the IT functions, in traditional organisations, are challenged daily by the increased expectations of CEOs and board members. Business leaders are accustomed to consumer technology standards, and want to innovate their business accordingly. Because of it, IT departments sometimes lag behind digital transformation programs. This chapter isolates and analyses five major changes factors impacting the IT practice: (**1**) *Pull-driven development*; (**2**) *Higher speed*; (**3**) *Technology democracy*; (**4**) *New suppliers ecosystem* and (**5**) *Social nature of digital*. These elements push for new vision, behaviour and leadership from CIOs and IT professionals. They also suggest a consequent adaptation of IT practices and strategies to overcome the change. Information Technology traditionally runs enterprise resources and assets, including its wealth of data. The commitment of the IT function is mandatory for succeeding in any meaningful, long term innovation journey. This chapter offers new options and views for those CIOs and IT professionals deciding to undertake a challenging change process.

The views expressed in the paper are those of the author and do not necessarily reflect those of his employer company.

D. Rizzo (✉)
Autogrill, Milan, Italy
e-mail: drizzo27@gmail.com

1 Why Digital Is Disrupting the Traditional IT Practice

The rise of digital is acting as a disrupting economic factor for most business and industries in mature economies worldwide. Digital native organisations (Amazon, Google, Uber, AirBnB etc. being the most common examples) have reshaped a growing number of markets through innovative offers and business models embedding Information Technology at its very core. In facts, Information Technology turns innovative business models into real ones by connecting customers with products and service providers, using computing algorithms, simplifying tasks once considered complex, etc. New digital players expand their business by eroding market shares, customer base and profitability to traditional players. In a growing number of cases, entire sectors such as libraries, bookshops, hotels and public transportation have been reshaped. Digital leaders dominates new markets where Information Technology is promoted as a core competing factor. Now, is this just the "new era" that most CIOs—historically promoting Information Technology as a competing factor within their organisation—have been fighting (and asking for funds) for many years? Are they, and their IT departments, gaining more power across such business technology intensive times? Are they, and their IT organisations, heading the digital transformation within their organisation, helping their business compete from a strengthened position? Although logically consequential, the answer to such inquiries is negative in many cases (Cox 2014). A recent survey by The Economist magazine titled "The disruption of the IT department" concludes that "...current wave of technological innovation has more profound implications for the IT department than any other function".[1] In fact, a closer observation on how established traditional organisations react to the digital challenge shows that, in many cases, digital transformation strategy is often delegated to or executed by newly established Chief Digital Officers, whose background is dominated by Marketing and sales experience.[2] CIOs and IT departments are often not leading such change initiatives. In a more limited number of cases, they do not even play relevant roles in the digital programmes. Only 3% of CEOs consider CIOs as a source of digital and IT-related business ideas, according to Gartner reports.[3] In fact, the digital factor acts as a powerful force by discontinuing the way organisations have historically dealt with Information Technology. Digital moves IT from a hidden role supporting ancillary, back office functions (Administration, payroll, etc.) to a core compelling factor. Digital Technology is more and more bonded with business model and revenue generation. It brokers a growing portion of the customer's experience. It generates and embeds product or service differentiation, sometimes impacting an organisation's survival itself. At a given extent,

---

[1]See also http://transformingbusiness.economist.com/the-disruption-of-the-it-department/.

[2]See "The 2015 Chief Digital Officer Study" published by PWC, dec 13, 2015, according to which 13% of companies in Europe do have a CDO.

[3]See: "The CEO perspective 2016: how CIO Should respond" by Mark Raskino—Barcellona Gartner Symposium 2016.

Information Technology could be redefined into *Digital* as it moves away from a complementary, technical matter dominated and understood by insiders (and often left to them) into a core factor enabling mass markets competition. Mobile, Cloud, IOT, Social, Analytical and Cognitive are the typical technology set developed to fight into digital markets. In this perspective, Information Technology stands for Digital in the way that car manufacturing stands for Motorsport, where competition is the differentiating factor for both cases. Many CIOs and IT teams are then exposed to new tensions and challenges as digital is escalated into a top priority within their organisations, and technology capabilities need to be incorporated into new products, new services or processes, or it is simply used to understand and attract customers or a competitor's strategies.

Further to this, the accelerated maturation of digital technologies,—primarily the cloud—pushes traditional internal IT practices, such as infrastructure management, towards obsolescence and potential irrelevance. Easy accessibility of 'pay per use' digital services on the consumer market (e.g. Infrastructure *"as a Service"* provided by players like Amazon, Google and others) makes the 'buy' alternative a cheaper, quicker and better option than traditional 'in house' IT infrastructure management (the 'make' option), where scale factor and technology trends make these costs and quality gaps irreversible. Any defensive reaction eventually adopted by IT management would certainly lead to a long term failure.

We have identified five elements driven by digital and consumer technology practice which are redefining Enterprise IT (see Fig. 1).

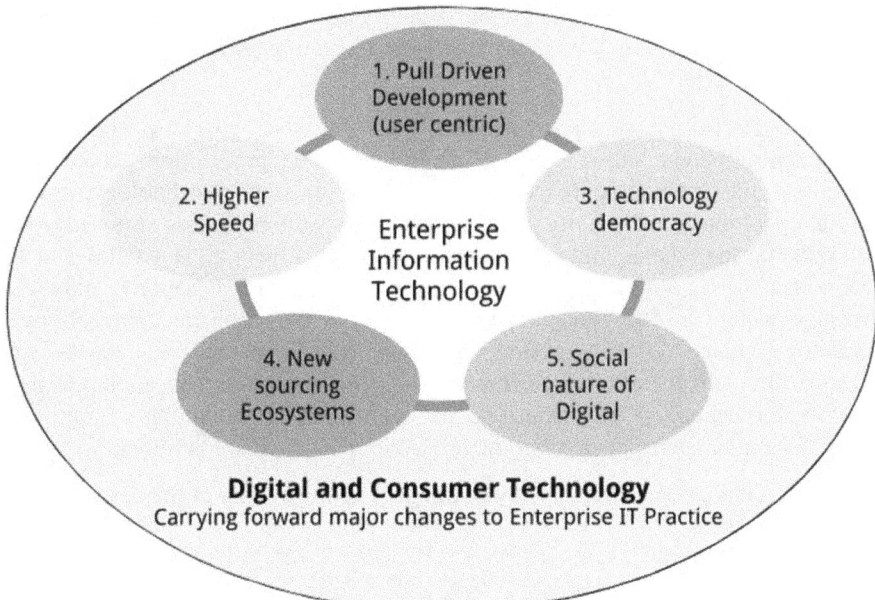

Fig. 1 Digital and consumer technology drivers changing Enterprise IT

1. User Centric design (from Push to Pull driven development)
2. Demand for higher speed
3. Technology democracy
4. A new sourcing ecosystem
5. The social nature of digital.

Each of them will be dealt with in the following paragraphs as a conceptual reference to promote higher awareness both for IT Professionals and Business decision makers, in order to feed views on possible future developments of the IT function into an organisation's strategic asset.

IT leaders and professionals are required to undertake unprecedented changes in those sectors where digital has emerged as a critical competing factor, and the board and senior management are asking for higher speed on technology related matters, so creating new interactions—and professional competition too—around digital technology governance.

But, on the other hand, the availability of Information Technology culture is crucial for organisations in order to:

- Match scalability, operational continuity and security requirements, sooner or later compelling elements of any serious transformation program
- Assure connection between digital innovation projects and core corporate assets and resources, normally run through traditional information systems. This condition reduces risks of creating "digital innovation islands" instead of a true digital transformation
- Fully exploit data, and sustain a data-driven strategy and analytical culture
- Nurture continuous innovation, as deeper understanding of technologies is crucial for upgrading business models, and creating differentiation and sustainability upon it.

CIOs and IT people can play relevant roles in those contexts where these requirements become clear, by redefining their value proposition accordingly. Modelling and systemic approach, design attitude, passion in technology, project culture, operational and security professionalism, as well as process knowledge are still valuable ingredients and typical skill assets of IT people. But CIOs and IT People also need to send new, clear messages to their organisations; leave old positions which are no longer sustainable, like a monopolistic approach over technology choices or conservative reactions to make-or-buy challenges—e.g. Infrastructure delivery or service management. Focus on true organisation priorities and consider technology as a means, not as its goal. Finally, obtain inspiration from successful digital start-uppers, who never seem to lack three conditions.

– Be obsessed by sustainability, by putting the business case at the center of any project
– Stay risk oriented
– Be intimate with software.

2 From Push to Pull Driven Development: Straight Towards a User Centric Design

The digital era has largely redefined market rules by shifting power in the hands of individuals. Access to a broad range of information, products, education, social relations is by now a vested option for million of connected people who can access the internet through their personal devices and apps, evaluate alternatives and expect to make their free choices in an open, virtual space, far beyond their formerly established constraints. Booking flights through one or more booking platforms, aggregating thousands of offers, options and prices is just one example of market pull approach, which also shows how the digital paradigm used by low cost airlines as they entered the market, and later on by web booking platforms has worked by focusing on travellers, handing over booking operations straight to them, while cutting old brokers (travel agencies) out of the game. New entrants have considered final users as the most powerful driving force of change, and they have used digital technology to focus on their needs (convenience, low prices, simplification of travel planning) delivering powerful and exciting customer interfaces, a better experience and, finally, lower prices (Ismail et al. 2014). There are more, less visible digital technologies increasingly being used now by airlines to optimise airplane load factor, pricing, and other key business parameters too, in a context of still complex, physical operations and capacity-driven industry structure. Nevertheless new, pure digital players (Expedia, eDreams, Kayak, among others) who use technology to support customer centric strategy have gained strategic market niches within a complex global industry. By doing that, they also set new, higher standards of direct interaction with customers and new higher expectations which are now spread across the whole industry.

In general, more free choices made by a large mass of individual users have determined the growth of most of digital and consumer technology markets, from gaming to smartphones, up to the web space. Success was, in fact, achieved by organizations resolutely determined to gain customers' attention and complicity through their own technologies embedding and developing value propositions (and finally their business models too).

A search on the internet, as well as the download of an app to play personal favourite music on a smartphone are gestures deliberately decided by individuals who feel their needs paid out by a grateful end to end experience; just by doing this, they are determining the success of its creators. The use of information Technology has consequently overtaken its previous limits—just consider the capacity to treat, store and retrieve huge volumes of unstructured data as example—right in the digital space, because of its "Pull" drive. It's no surprise if recent, amazing steps ahead in computer human interfaces have been mostly driven by consumer technology companies like Apple, Google or even by much smaller firms like Instagram, Splice or Pandora, instead of established players operating in the B2B Information Technology market. Higher intimacy with customers is still piloting computer applications far beyond the old borders of personal privacy when considering fitness and health app providers, who treat sensitive personal data with the full consent and collaboration of millions of individuals, whose life is under tracking by someone else.

Enterprise Information Technology instead, has historically selected and funded its projects through a top-down approach. It has delivered cost and control improvements through the massive automation and standardisation of repeatable processes like payroll, administration and cash operations. The differentiation of processes, proliferation and personalisation of use cases has been treated as a driver of inefficiency and cost increase. Exceptions to standardisation have been reduced up to the optimal trade-off and—still more relevant—the user's adoption has been achieved through hierarchical levers (the adoption of a new stock management procedure has never been a choice for warehouse clerks). Marginal investments over more "user friendly" interfaces have been developed as a complimentary lever to reduce change attrition, being sometimes considered as a "tax to technical ignorance" to evade wherever possible.

Information Technology professionals need to be aware of their cultural legacy. They must open to a different, much more radical User Centric approach. Digital technology democracy is moving power into the hands of technology users; this is happening within enterprises and institutions too. IT people need to realise that the choices and behaviour of individual users will decide who will survive in the market, and who will not.

3 From "Getting It Right" to "Failing Fast": Managing Innovation Through Technology Speed

Digital experience is normally coupled with a dramatic acceleration of IT artefact production times, when compared with equivalent processes of traditional enterprise IT practices. Consider as an example the average frequency of version upgrade of apps on mobile stores, often delivering two or more releases per month, where more traditional enterprise IT Systems normally undergo the same frequencies of changes in a year. Similar examples take place in the web space too, where popular portals supporting complex information services like Google or Facebook are subject to almost continuous releases (Beta permanent versioning).

Speed increases not only apply to software change management processes. The elapsed time required to release a new digital product version from business requirements to first product availability is constrained "by design" into typically 1–4 weeks within Scrum sprint frameworks (Sutherland 2014), while more traditional, "waterfall" enterprise IT projects could last months or years. Compression of time is achieved by reducing business analysis through highly interactive and focused sessions, freezing versions until the next release, fast availability of prototypes which anticipate customer interaction impacts (e.g. by using design prototyping public platforms like InVision), implementation through web pay-per-use infrastructure and software components, high reuse of software and public APIs, reviewed testing policies supported by automation and external sourcing. The growing spread of Agile and Devops disciplines support such changes on a methodological perspective, while an increasing number of available public platforms (e.g. Github or Amazon AWS) offer easy, wide and democratic access to productivity and speed improvements. More interestingly, such speed improvements are not associated with any quality compromises; on the contrary, they seem to offer a new, effective way to control risk for those investments exposed to highly volatile and uncertain environments (e.g. products innovating customer's experience), by reducing time and cost required to check out project assumptions and consequently drive and adjust further developments and investments. The alternative going for higher investments and longer time spent in the analysis and design phases—typical of waterfall methods, used in many IT Projects—may turn into a late and expensive discovery of discrepancies between hypothesis and reality. Deep business implications of such relationships between speed and frequency of artefact development cycles and successful investment management, mostly in the field of digital products development, have been widely described by Eric Ries in "The Lean Startup" (Ries 2011).

CIOs are now facing the challenge of re-adapting the governance of technology investments accordingly. IT investment decision making in the '90s and early 2000, was crafted around a deterministic, predictable approach in many IT Departments. The past era of technology standardization (i.e. erp, scm implementation) requested a top-down governance approach to select investments and drive change downwards, focusing the whole organization on specific and predetermined KPI improvement goals. IT Project methodology was based on best practices (e.g. PMI, PRINCE2) mostly adopting a sequential and deterministic paradigm. Technology itself was not even ready to offer ways to cut implementation times or reduce project cost thresholds (e.g. by providing ready to use resources on the cloud). But a much wider availability of ready-to-use digital resources is now driving IT into new governance paradigms, where a traditional approach must be integrated with a more empirical and experimental one. Improving the speed and the number of available options (flexibility) within a project reveals it to be a much a more effective strategy than asking for time to take a single "silver bullet" shooting option. Project strategy dynamically emerges as a result of a number of reiterated propositions, when prototype outcomes are analysed, interpreted and adjusted in strict conjunction with project objectives. A growing availability of contextual data showing relationships between the digital artefacts and their users (e.g. think about transaction logs, in store camera shootings, store wifi reports when developing a self service digital kiosk within a restaurant), together with a frequent collection of user insights shortens feedback loops and controls project risks. However an "engineering" and sequential mindset of some IT people may sometimes act as a barrier to this shift, when IT people expect completeness of requirements and clarity of directions in contexts where clarity may only be around goals, expected outcomes and constraints (speed being among these).

4 The Era of Technology Democracy: Rise and Establishment of the Digital Connected Community

The advent of Salesforces.com in the early 2000s, and its fast-growing dominance[4] in the Enterprise CRM Software market achieved in less than a decade to the detriment of giant enterprise software incumbents is probably one of the most disrupting events in Enterprise IT practice in recent times, as well as a didactic case of the direct impact of Digital on Enterprise Information Systems. The CRM software market before Salesforce was characterized by licensed products run on customer's premises. Its selection and availability for use required projects involving IT professionals involved in technical activities, supporting Sales and Marketing departments often through third party system integrators, finally delivering customised single instances of the product. Salesforces entrance into the market has marked a different—and much simpler technology adoption pattern. Software is supplied as a service available for use through a powerful public cloud platform, simply accessible through the internet. Infrastructural IT activities required to run it are then limited to modest network security access configurations. The success of the platform was consequently decreed by its users community driving its selection process—Sales and Marketing professionals—appreciating the direct focus on customer engagement and marketing targets results, attractive interface, users self configuration capabilities, open connections to social networks, a wide offer of built-in apps and a rich software marketplace fed by independent third parties, and many more elements, normally associated with a digital consumer's experience. The whole of it, built upon a strong and solid technology core. Out of the Salesforces example, similar cases have marked the course of more and more enterprise software segments like HR Performance Management, Purchasing market platforms and others. Sometimes, such adoptions have taken place as a hidden process to IT organisations, being originally classified as "shadow IT". Such closed loops between users (consumers) and technology providers, lead to the disintermediation of IT organisations and create new tensions and challenges on the ability of IT to contribute to business technology design and value generation. More in general, the availability of cloud applications and pay-per-use technologies is shifting technology choices in the hands of final users out of deputed technical organisations. This trend reverts traditional enterprise top-down technology governance into a bottom-up process (see Fig. 2). Similar patterns have already emerged with enterprise personal devices, where company provisioned PCs and mobile phones have mostly given way to "Bring Your Own Device" policies driven by users' ambitions to decide on their technology by themselves, or in the area of customer care and support, where people are addressing their complaints or sharing their request for support on their preferred

---

[4]Salesforce Leads the Worldwide Enterprise CRM 2015 market revenue share with 19,7%, ahead of SAP (10, 2%) Oracle (7, 8) and Microsoft (4, 3),—See Gartner's "Market Share Analysis: Customer Relationship Management Software, Worldwide, 2015—Published 12 may, 2016.

social networks like Twitter, Facebook or others, instead of queuing their complaints within an official and often frustrating IVR Call Center. Security itself is evolving accordingly, moving away from pure perimetral to logical access control, integrated by big data, real time analytics and other technologies, integrating social login credentials and authentications across different platforms, enabling more open architectures, where private networks and on premise datacenters are no longer the center of the universe. The interpretation of these trends, trade-offs and side effects on security, scalability and integration is generating different answers from different CIOs, depending on Industry characteristics, company culture and CIOs personal attitudes. Nevertheless, even within less exposed environments—e.g. sectors where security is predominant—It is becoming clear that CIO's full monopoly over technology is no longer sustainable, and some degree of opening to technology democracy must be found. By giving up constraints, past CIO's gatekeeper's power shifts away (the keys of its data center are no longer a source of power) and the battle for business relevancy must be fought on different fields. Integrating technology components and external services within a consistent business model as well as making information available and accessible across the different stages of business processes are still complex tasks which IT people should take care to design and develop, out of technology monopolies in a new relationship with their peers, looking for the right balance between challenge and collaboration (Fig. 2).

Influence on technology choices shifting from institutions to individuals

Fig. 2 Influence on technology choices

5 New Technology Ecosystems: Emerging Players Redefine Enterprise IT Sourcing Practices

The industry of Information Technology has generated a rich offspring of market champions across its relatively short history in the second half of 20th century. Corporations like At&T, IBM, HP or Microsoft have not only dominated their own markets, but also covered absolutely the top positions of global financial ranks during the early 2000s. They formed the peak of the IT suppliers pyramid, ranging from large and global to small and local firms, supplying Hardware, Software, Technology services and Telecommunications to companies and institutions using those building blocks as components of their information systems. Information Technology departments and CIOs were part of this ecosystem as top buyers and business integrators for years.

The following digital outbreak across the 2000s did not actually discontinue main core computing and network technologies, still feeding lower lever technology manufacturers like Cisco and Intel with new growth opportunities. However, the outburst of digital economy was written and driven by different subjects. In most cases it concerned new start-ups who were able to generate innovating offers designed for consumer driven markets. They run capitally intensive and easily scalable business models, mostly funded by venture capitals. With a few exceptions (Apple *in primis*) a new born generation of subjects like Google, Amazon or Facebook dominate today's digital markets. These new players have not just sold technology. Instead, they have monetised its potentials by creating innovative business models run through technology. They have invented platform economies by connecting consumers and providers in a wider ecosystem of different stakeholders, each of them contributing to enriching the customer experience and improving the platform value itself. Their revenue models have shifted from sales of technology components (e.g. hardware or software licenses) to the monetisation of their big transactions volumes (e.g. marketing personal advertisements or e-commerce transaction markups). Although their original business model focus was set far away from Enterprise IT, nevertheless some of their innovating technologies (e.g. Amazon Web Services, Google for work, or Facebook's CRM APIs) have been adopted by a growing number of enterprises and institutions as part of their information system architecture. These new offers, often cheaper and easier than the traditional enterprise IT alternatives, have brought massive disruption to the enterprise B2B IT market and to traditional IT providers.

The new wave of digital players is now heading global financial ranks of market capitalisation, in some cases replacing older technology corporations; and similarly to the pre digital age, a much larger number of small to medium size companies is breaking into a highly dynamic and volatile market ranging from bio to fintech, industrial IOT, education, gaming, food, automotive and much more. It represents a growing segment of economy highly appetizing to capital markets, but also attractive to traditional businesses and established institutions who consider them as a powerful and credited source of innovation and an opportunity to transform their organisations into a digitally connected business environment. However, differently from pre digital tech companies, IT departments are no longer "natural buyers" of such players, who deal with consumers or business decision makers. Conversely, even more traditional IT players are modifying their products and market strategy, consequently targeting the same business interlocutors instead of, or in addition to CIOs. This change is nothing but the further evidence of a new and more connected framework of relationships between technology, business and institutions.

The ecosystem of digital suppliers is then discontinuing the traditional IT sourcing framework and work relationship context, creating new tensions and challenges to the IT function. CIOs need to understand and adapt to such a changed environment, moving beyond controlled sourcing relationships purely focused on technology, towards broader collaboration contexts made of business connected supply agents, thus maintaining a direct, external key source of innovation and personal development. Process design, sustainability and scalability of digital business is still dependent on technology, and a higher knowledge around its domains is required when making technology acquisition choices.

CIOs also need to develop a new language and value proposition to interact with new subjects; build different sourcing opportunities in a much more turbulent supply environment, through continuous scouting of new players; set strong collaboration protocols with business functions, in order to avoid overlaps and boost a multidisciplinary approach across sourcing processes; practice new unusual recruiting channels in order to make new technology resources and opportunities available to their organisation community instead of just waiting for the natural, digital evolution of their old and trusted delivery partners.

6 The Social Nature of Digital: The Information Control Leaves Way to the "Sharing Economy"

Digital and Social are keywords which have been strongly paired since the mid 2000s, when social networks were established amongst the most pervading innovations generated in the digital era. By the time startups like MySpace or Flickr, and then Facebook had launched their revolutionary social interactive web services, social capabilities of digital artifacts have grown intensively and pervasively. Today, some degree of social interaction is part of the experience of almost any digital service, from music to design, from ecommerce to car sharing. The social applications of digital, enabled by cheap access to the internet and innovating applications have already changed the patterns of interaction of individuals and groups of modern societies not only at a business level, but also in politics, private relations, personal and medical care, no-profit associations and more. Gartner has included Social as one of the four convergent, disruptive drivers of the Nexus of Forces, characterising the digital age. Digital social capabilities are providing individuals with unprecedented chances to share ideas, contents and freely interact with a potentially infinite number of other people, but they also bring new challenges both to privacy rights and organisational models in public and private sectors where availability, access and the spread of information configures power balances. They also bring extensive impacts and opportunities to CIOs and IT Organisations at three different levels:

- **Technology**—Until the mid 2000s structured data and relational databases were the most, if not the only technology available to organise information within an organisation's IT. At the times of ERPs, IT people dealt with SQL as the standard language to treat transactional data, just like businessmen did with English for business. So, to make information consumable, IT worked to transform unstructured information into structured data, by creating transactional interactions where information was constrained into data organised into relational databases. Social digital instead, has promoted unstructured data (texts, photos, videos, etc.) as the new information standard, as it is the natural content of human interactions. Technologies like NoSQL databases and CMS were further created to store, retrieve and treat unstructured data, barely usable before. These technologies are now part of a much larger tech set, ranging from voice recognition to digital assistants from video analytics to text semantics from chatbots to cognitive computing. Although generated in the social digital space, far away from enterprise IT culture,[5] these digital social technologies now knock back on Enterprise and CIO's doors, promising powerful applications and broad improvements in many fields.
- **Organisation**—The social capabilities of digital have created innovative opportunities also in the Business to Business segment, by leveraging on large communities of individuals to perform complex tasks historically assigned to trusted and credited subjects instead. The word *Crowd* has become a prefix to other words like *Crowdsourcing* or *Crowdfunding*, all of them suggesting alternative, network-centric relationships to outsource critical goals. Significant social innovations have more specifically affected also the processes and the organisation of Information Technology. Starting from software development, well before the digital era—and mostly out of IT departments—the Open Source communities have demonstrated the ability to develop quality software through a collaborative process committed to network-based virtual groups. More recently, several software related processes—such as testing (crowdtesting) or quality control—can be outsourced to network-based communities, where external developers or testers interact on assigned projects on platforms like Github or Testbirds. The effective use of such innovating outsourcing models requires changes in the organisational context of many Information Technology departments and redesign of the IT delivery value chain in a more open organisation framework.

---

[5]Even IBM's Watson cognitive answering system, whose applications range from healthcare to weather forecasts, gained its popularity by competing against humans in the US Television quiz show *Jeopardy!* in 2011.

- **Culture and behavior**—Huge opportunities enabled by digital social networking have encouraged behavioral and cultural changes on highly controversial topics like personal privacy and information control. In more and more cases, the perceived value of sharing information has overtaken concerns of doing it.[6] Beyond demographic and geographic differences, the attitude towards information sharing is subject to continuous challenges in the digital space. This is made evident by many social networks configuration settings, where public sharing is the default, and restrictions are an option—instead of the other way round. Wide access to information not only feeds concerns about personal privacy, but can also promote more democratic access to knowledge and information resources. Innovating companies like Google encourage their staff to a wider sharing of information. This behaviour can promote organizational transition from a hierarchical model to a more participatory, boosting creativity and growth. However, Information Technology departments and their CIOs have traditionally operated as information security agents within hierarchical organisations. They have historically operated perimetral security around Company information assets and personal privacy, protecting them all from theft or external attacks. Today, as cyber threats still require growing attention, more opportunities outside the company digital borders ask CIOs and their staff for a new balance between risk and opportunity management. Better trade-offs between growth and control require more selective policies and approaches (e.g. Improving information security by strengthening access instead of perimetral control). But also a cultural shift from IT People, who need to realise they can obtain opportunities, not only risks, from outsiders.

References

Ries E (2011) The lean startup: How today's entrepreneurs use continuous innovation to create radically successful businesses. Crown Business, New York

Cox I (2014) Disrupt IT: a new model for IT in the digital age. Axin, Ipswich

Ismail S, Malone MS, Van Geest Y (2014) Exponential organizations: Why new organizations are ten times better, faster, and cheaper than yours (and what to do about it). Diversion Books, New York

Sutherland J (2014) Scrum: the art of doing twice the work in half the time. Crown Business, New York

---

[6]See. "Facebook and Online Privacy: Attitudes, Behaviors, and Unintended Consequences" by Bernhard Debatin, Jennette P. Lovejoy, Ann-Kathrin Horn, Brittany N. Hughes—Journal of Computer.mediated Communication http://citeseerx.ist.psu.edu/viewdoc/download?doi=10.1.1.616.9669&rep=rep1&type=pdf

Future of the CIO: Towards an Enterpreneurial Role

Carlo Alberto Carnevale Maffè

Abstract CIOs can become co-entrepreneurs in their organizations by finding novel ways of addressing critical factors of scarcity: capital, labor and trust. By using modern software technologies and protocols, such as AI, smart contracts and blockchains, CIOs can redefine the nature of these three factors to digitally transform most of the value-added processes and make them available "as-a-service" to all relevant stakeholders. Furthermore, the function of "time," another critical factor for organizations, is also affected by the CIO's new role: beyond the mere logic of "deadline," time is articulated in timing (*"when"*), time-to (*"how soon"*), and committed time (*"for how long"*).

1 Future of the CIO

In undertaking new entrepreneurial roles, CIOs become co-responsible for the three main factors of scarcity in their organizations:

- **Capital**
- **Labor**
- **Trust**.

They must learn how to manage these three factors in a flexible, dynamic and efficient way, by making them **"software-defined,"** and by optimizing their usage along with the most critical cross-coordination function, **Time**.

Following Marc Andreessen's prophecy, a large chunk of CIOs' traditional world has been eaten by software. Most fixed-capital assets, from infrastructures to platforms, application packages to specific user devices, have become virtualized and are being made available "as-a-service," without requiring long-term capital commitments, dedicated and specialized labor, and contract-based trust. When transaction costs—which affect make-or-buy decisions of vertical integration,

---

C.A. Carnevale Maffè (✉)
Bocconi University School of Management, Milan, Italy
e-mail: carloalberto.carnevale@sdabocconi.it

along with the scope and scale of firms—fundamentally change because of technology, so must the entrepreneurial role of managers.

On software-defined Capital:
The CIO must learn to efficiently use capital, whether financial or intellectual, through the adoption of innovative solutions, from block-chains to digital rights management, API logic to security issues.

On software-defined Labor:
The CIO's job is to maximize the productivity of human labor by adopting digital technologies based on a vast range of solutions, from industrial automation to collaborative robotics, pegged services to distributed autonomous organizations.

On software-defined Trust:
The CIO shall be able to write new rules for the artificial production of trust and the enforcement of contracts; these rules will evolve from traditional legal texts written in natural language to the new form of "smart contracts" and artificial intelligence, using distributed "proof of work" and social rating platforms.

Being responsible for the three main factors of scarcity is increasingly pushing the CIO to interact with other business functions, both staff and line. **Capital** ties the CIO to the CFO and the Board of Directors. **Labor**, to HR and production/distribution functions. **Trust**, to the general counselor and to purchasing, marketing and sales.

The coordination function, i.e., **Time**, binds the CIO to the CEO and to general management roles.

1.1 Capital

Data is the New Money. As Thomas Gresham wrote in the 16th century, bad money always drives out good money. Data are the perfect "bad money" of current times, because they are abundant and liquid; they therefore increasingly constitute a viable replacement of "good," official legal tender. Indeed, real money has become a less efficient means of payment with respect to data, at least in the digital world. Ultimately, though, the conversion of data into real money (i.e., data monetization) is the big challenge of any new business model, which CIOs must learn to master and implement.

Software is making fixed capital effectively useless. The world's largest hotel chain, Airbnb, owns no hotels and hires no concierges. The fastest growing logistic corporation, Uber, owns no vehicles and hires no drivers. The biggest media company, Facebook, owns no TV channels and hires no journalists.

Today, software re-defines traditional forms of fixed capital in many ways: directly, through data as intangible collateral and virtual currencies for payments; indirectly, with virtualisation and on-demand access to physical assets and seamless management of digital intellectual property rights through APIs (Application Programming Interfaces).

CIOs have come a long way in transforming their own traditional portfolio of fixed assets (data centers, device fleets, permanent software licenses, etc.) into virtualized resources that require much less long-term capital commitment—or none at all-, thus impacting balance sheets as well as cost structure. Invested capital de-leveraging, based on software-driven virtualization and access, must now be pursued for a much broader range of assets, including the company's, the partners', and the customers'. Application Programming Interfaces help organizations do soby allowing CIOs to safely 'expose' the functionality of their internal applications or services to the outside world in a controlled manner. With APIs, asset capabilities are efficiently virtualized, decoupling applications or service implementations from their final users, and providing building blocks for third parties to create additional services; these services can be remunerated—whether with monetary or non-monetary exchanges—via the same APIs, through a fair share of the total added value.

For the future provision and exchange of financial capital, including loans, credit and payments, CIOs need to be ready to take part in the design and implementation of new block-chains for distributed certification of money transfers, and in general for co-operative, digital "proof of work" of any financial transaction.

1.2 Labor

We are still far away from the much-trumpeted "Technological Singularity"—the moment when machines will be as "intelligent" as humans—or maybe even more intelligent as, unlike humans, they wouldn't be bound by biological constraints and because they would enjoy recursive self-improvement, evolving their own capabilities in an exponential way. But the job of CIOs has always been to extract maximum productivity from human labor through information technologies; they must keep on leveraging and enhancing human capabilities by transferring the more mundane and repetitive human tasks to computers, as well as progressing towards machine learning and software-based inferential skills. The controversial empirical results of organizational research on the actual productivity gains resulting from computers need to be challenged with unquestionable outcomes in terms of potential value added per head. After all, HR is the only corporate function that appears to be stuck in the nineteenth century of analog management disciplines. CIOs mustengage in experimenting new forms of division and coordination of labor, both internally and externally, dealing with innovative solutions for managing incentives, substitution and complementarity.

The current organization of labor, both physical and intellectual, shall be re-divided between humans and machines, creating hybrid solutions within complex Cyber-Physical Systems (CPSs). CPSs can be described as "systems of collaborating computational elements controlling physical entities". CPSs become software-defined "co-workers" of humans with smart factories and advanced production environments, where manufacturing units leverage information under human supervision, progressively learning to behave autonomously.

But CIOs will have to learn to organize and coordinate not only the labor of theircolleagues (whether humans or robots...), but also the tasks performed by external subjects, in particular by final customers. Operating on digital coordination platforms, in the so-called "gig economy," matches are created between a pool of available, skilled workers and user-defined tasks, on-demand and in real time. For every customer's task request, a formal digital contract with available workers is activated, with the obligation to provide services (hospitality, car riding, home delivery, etc.) and access to their asset (a car, a house, etc.) with the on-demand company in exchange for a commission of the management of the platform's processes.

CIOs will therefore move their focus from Data Centers to Data Borders: they become "security officers" patrolling the perimeters of organizations, constantly moving the borders across new equilibria of transaction costs. Their goal will be torelentlessly pursue the digital/robotic complementation (and/or substitution) of human labor with software-defined labor.

1.3 Trust

In a global digital world where national authorities and local applicable laws are sometimes unacceptably slow and possibly ineffective in exercising the power of contract enforcement, technology and organizational innovation are providing different solutions for the critical challenge of establishing trust among trading parties that need real-time, undisputable and non-refutable contracts and their consequent execution. The internet and its collaborative platforms, in this respect, are no longer just a technology infrastructure, but a new economic institution, probably the only truly global institution of modern times. In this context, CIOs will become underwriters of "smart contracts" to create trust, not in the legal jurisdiction of local authorities, but through the aggregation of members in coordinated block-chains. These software protocols can facilitate, verify and enforce both negotiation and execution of a contract, entirely in digital form, even between unknown parties.

The advantages of using smart contracts over traditional forms of textual agreements, defined by economic literature as "incomplete contracts," are numerous. In practice, incomplete contracts cannot include provisions for every possible contingency. At the time of the agreement, future and/or unexpected contingencies may not even be describable. In that context, the parties would prefer to engage in renegotiation later on in their relationship, generating the so-called "hold-up" problem. The probability of future renegotiation, in fact, reduces the incentives to commit to relationship-specific investments. Conversely, smart contracts in their most essential form can be defined as "complete" contracts. They can be made partially or fully self-executing, self-enforcing, or both. They do not require re-negotiations, nor transaction-specific investments. Smart contracts can provide better security performance than traditional contract law and reduce most transaction costs associated with negotiating and enforcing contracts. In reality, their combination and aggregation may constitute an "incomplete" contract. But standardizing the components of a general agreement through simple, interoperable, non-repudiable smart contracts can transform an entire business model. Any modern CIO must stay ahead of this evolution, by learning to be a "notary "of smart contracts.

2 The CIO as Market Maker

CIOs will need to learn the art of market makers of information exchanges, or—better—creators of "Fiat Markets" made by information exchanges. In these new markets:

(a) **Data** become *Money*
(b) **Communications** become *Contracts*
(c) **Conversations** become *Commerce*.

CIOs will therefore become increasingly involved in Economics of Data (EoD). They will be in the best position to design the multi-sided market insofar as they may understand—better than other functions in the firm—the management of information externalities, both positive and negative, as well as the options for cross-subsidization of participating subjects. By having CIOs play an important role in EoD, firms can identify potential users and providers of data to establish a multi-sided market structure, building an ecosystem of partners and defining shared data asset platforms to be used by all players. The potential for value creation derives from data sharing, both in a raw state and enriched with additional sources and structures; this data sharing provides benefits to multiple parties that, in turn, extract value for their own business context. As a further sophistication, data semantics can leverage feedback from participants to extract further value and meaning. This incremental insight is then shared and distributed across the platform.

In their new role as market makers, CIOs will necessarily be influenced by agency theory. Their new (and old) job will be to overcome the information asymmetries and solve the classic, theoretical problems of hidden characteristics, hidden actions and information, and hidden intentions. Such problems are at the basis of the contract theory of the firm and the transaction costs approach.

In economic literature, information asymmetry is articulated in three main elements:

Hidden characteristics: Certain features of the agent and/or of the goods being exchanged are not known before the contract is made. Hidden characteristics are features about one side of a transaction that are known by one trading subject and that the other side would like to know but does not. A hidden characteristic is intrinsic to an agent, and it's relevant when facing the risks of "adverse selection". With the use of Big Data, analytics and pervasive profiling, CIOs can make the characteristics of subjects and/or goods and services involved in contracts less and less uncertain.

Hidden action and hidden information: after the definition of a contract, not all of the actors' actions can be observed (hidden actions). E whenobserved, qualities of the relevant action may not be easily determined (hidden information). A hidden action is potentially voluntary, therefore paves the way for opportunism and moral hazard. Social rating, distributed proof of work and smart contracts organized in block-chains can be powerful tools for reducing the risk of hidden actions and information.

Hidden intention: even before a contract is made, actors might be observable, but their real intentions cannot be known. Intentions can be revealed—or faked—through "signaling," but are prone to misunderstanding and (un)intentional distortion. Today, with systematic behavioral profiling, machine learning and artificial intelligence, CIOs have powerful tools for the analysis of "signalling": intentions of various players can somehow be to a certain extent, anticipated.

In the future, CIOs will be able to overcome not only information asymmetries, but also "organization asymmetries" between supply and demand, by designing processes that implicate customer participation in value-added business processes.

Organization asymmetries between demand and supply are related to the different levels of capital, labor and trust. Firms, on the supply side, used to enjoy and benefit from a significant level of organization asymmetry with respect to disperse, individual customers, especially when dealing with final consumers. CIOs will have the privilege—and the duty—of becoming makers of new markets, where both the supply and the demand side will overcome information and organization asymmetries, targeting a higher and more stable equilibrium of value, away from zero-sum games, towards positive-sum games for all players in the ecosystem.

3 The Fast CIO and The Strategic Management of Time

In order to achieve a better equilibrium, the three major factors of scarcity (capital, labor, trust) must be managed in a synchronized manner with the most important coordination function, i.e., Time.

After years of justifying their decisions with financially approximate variables such as the alleged "cost" of their budgets, CIOs are discovering that the real critical issue in ICT is **time**. Time is indeed the decisive, influential factor on the value of special "experience goods," such as digital products & services, where cost and overall value are only known ex post. In these cases, the annual cost accounting becomes a subjective opinion rather than objective evidence and, as such, is affected by political decisions, financial tricks (e.g., outsourcing decisions reduced to mere financial swap between fixed assets and rental payments), and simple economic and organizational ignorance. Moreover, it's difficult to measure the opportunity costs of ICT in certain cases. In choosing a particular type of architecture, for example, the evaluation of short-term alternatives remains somehow arbitrary, due to the turbulence of technological life-cycles, that can make any investment a sunk cost in a very short time.

By acknowledging that the cost of ICT is a dependent variable of time, CIOs are faced with its multidimensional nature; for a CIO, the relevant elements of time are the following:

- timing ("*when*");
- time-timpaco ("*how soon*");
- committed time ("*for how long*").

Technological evolution is constantly changing the nature of these three major constraints for CIOs: "timing" is often imposed by consumerization or "pull" adoption by customers. "Time-to" is influenced by availability of APIs and DevOps; "Committed time" becomes opportunistic with cloud-based services, but it often ends up out of management control.

The "**timing**" has to do with the timeliness of decisions, and takes into account both external and internal timing constraints. On the external front, the CIO has learned the hard way that excessive frontloading of technological choices, far from creating a first mover advantage, becomes an expensive exercise of becoming the test bed for immature technologies; these failed tests often benefit only unscrupulous vendors or more prudent competitors. The regulatory requirements are another external constraint; they often impose less discretionary timing, but offer, in return, important windows of opportunity to force decisions that impact the entire organization, thanks to forced discipline. Internally, the correct timing requires the CIO to be aware of the need to move in synch with such organizational constraints as the maturity of the required skills and the development of related processes. Failures of many innovation projects are often not the result of excessive technological advances, but rather by internal immaturity.

"**Time-to**" is the next logical dimension and relates to the time lag between the decision and the actual implementation. This dimension also has constraints on both the external front and the internal front, and the CIO must address significant trade-offs. The systematic use of external consultants for the execution of ICT projects can gradually erode internal capabilities of learning and cost control; the adoption of a standardized SW package, for example, can reduce uncertainty, but has the effect of entrusting the control of both the "timing" and the "time-to" to a third party vendor: examples are the forced passivity about release dates and new versions of operating systems, not to mention cloud services' SLAs. On the internal front, with decisions relating to the "time-to," CIOs must manage the expectations of the various stakeholders involved and the synchronism with other interdependent projects.

The "**committed time**" concerns the implicit and often unconscious commitment that the CIO takes regarding the future of a certain technological choice. It is the root cause of the phenomenon of legacy systems and results in greater switching cost and, therefore, the possible effects of lock-in with respect to certain vendors. In the ICT budget, the "committed time" erodes room for discretionary choices and stiffens the flexibility of expenditures.

From the managerial point of view, these relevant dimensions of time can be lessened by adopting different attitudes, which we can define as "passive", "opportunistic" or "strategic" management of time.

Adopting a passive management of time often reduces the role of CIOs to a hostage of exogenous forces: they become dependent variables from other organizational roles, dragged by vendors or driven by the latest technological fashions. An opportunistic time management is somehow justifiable on specific occasions, such as the approach of a relevant regulatory compliance, to exploit forced discipline and to obtain significant investment budgets and spending authority. Only with a strategic management of time, however, is the CIO able to rule all the relevant dimensions listed above, retaining the right to control the levers of timing, time-to and committed time. The CIO, after learning the necessary virtues of thrift and integrating strategic business processes, must become "fast" in every sense of the term, that is timely, quick and flexible.

CIO's: Drivers or Followers of Digital Transformation?

Giancarlo Capitani

Abstract It is increasingly clear that Digital Transformation acts as an enabler of structural innovation of company processes and products, services and go-to-market models and that it does so through the appropriate use of new digital technologies. In this sense it appears as a contextual transformation of a company's business strategies and technology infrastructure. CIOs are correspondingly required to transform their own role by offering themselves as leaders of Digital Transformation. However, if Digital Transformation is a great systemic project, consisting of a long process guided by a master plan, what factors should direct it if it is to succeed? And what tasks should the CIO perform to bring this process to a successful conclusion? In this context, CIOs find themselves at a crossroads requiring them to redefine their role in the immediate and medium term perspective. The results of a survey conducted in the first half of 2016 on a panel of Italian CIOs (CIO Survey 2016) provide us with an answer to these questions. Specifically, the Survey reveals 10 lessons learned that in the experience of the CIOs interviewed are key requirements for the successful completion of a Digital Transformation project.

1 The Reality of Digital Transformation in Companies and the Role of the CIO

Digital Transformation is currently the prime strategic goal of European companies. It is also backed by their respective governments, which through major public programmes, such as Industry 4.0 and Horizon 2020, seek to stimulate and govern the digitisation of their countries.

However, it is clear that at present there are no established and unambiguous models and roadmaps at enterprise or national level to which we can refer to achieve this ambitious goal.

G. Capitani (✉)
Politecnico di Milano, Milan, Italy
e-mail: capitani@netconsultingcube.com

The consequence of this is that even with a widespread belief that Digital Transformation is a real tool for the structural innovation of companies and the modernisation of a country's economic system, companies and public administrations follow an individual path, trying to draw the greatest possible benefits from it.

This transitional scenario poses many questions that do not always meet definite answers, starting from an understanding of what the current and future role of the CIO might be and also, more radically, whether the CIO will play a role on the path towards Digital Enterprise.

2 The Results of a Survey

In answering these questions, we could find it useful, for example, to consider the results of an Italian survey that has been conducted for about 10 years (CIO Survey) by NetConsulting Cube (2016). This survey aims to monitor the projects, perceptions and opinions of a panel of 100 CIOs operating in major Italian companies but not in Public Administration.

The distribution of the CIOs chosen on the panel represents their companies' respective industries in accordance with their relative importance within the Italian economic system (Fig. 1).

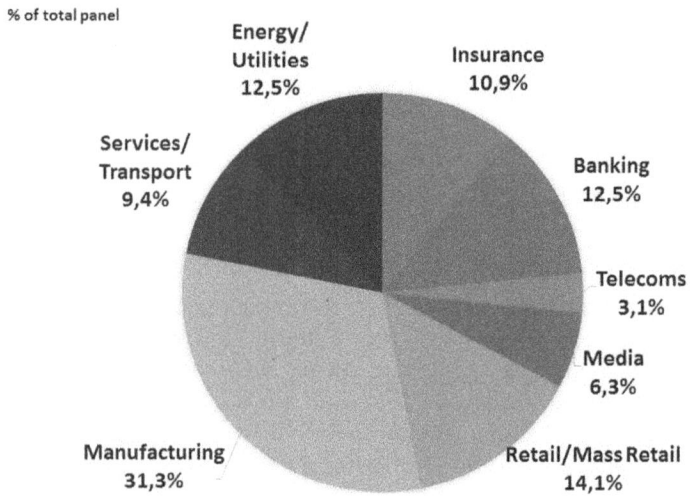

Fig. 1 Composition of the panel. *Source* CIO Survey 2016

The Survey focused on the following aspects:

- How CIOs and their organisations are coping with Digital Disruption—what impact it is having on ICT resources/skills, infrastructure, applications, and governance
- Digital workshops utilising new technologies—IoT, Cloud Computing, Big Data, Mobile, Social Media—through which they are implementing Digital Transformation
- The Impact of Digital Transformation on the ways in which ICT interacts with other company functions, on the governance of innovation in the company
- The dynamics of ICT expenditure and investments
- The changes in sourcing policies and the methods of interaction with external partners.

Using the results of the last edition of this survey (CIO Survey 2016) we can translate the views of the CIOs into at least 10 lessons learned from their everyday experience.

3 Lesson 1

3.1 CIO Vision: Digital Transformation Is not just Technological Innovation

A large proportion of the companies interviewed have activated digital workshops and are accelerating their investments in projects in areas such as mobile (88%), cloud computing (76%), big data (70%), IoT (42%). However, while until two years ago CIOs believed that the optimal model of Digital Transformation was the synergistic integration of these project areas in order to innovate a company's technology infrastructure, today their perspective has changed and is increasingly based on a business and strategy driver vision Uhl (2014).

The new vision sees Digital Transformation as a systemic innovation based on three pillars (Fig. 2):

(a) Customer Experience, that is the creation of an interactive relationship with the digital customer where the factor of attraction and retention is experiential and emotional in nature in support of the service quality of the products offered. Technology infrastructure based on digital paradigms must be able to provide customers with omnichannel, physical and virtual access to the services
(b) The reorganisation of internal processes in order to create a form of cooperation, interaction and exchange of data and information between different company functions (Smart Working) also extending these new interactive and transactional methods to external partners

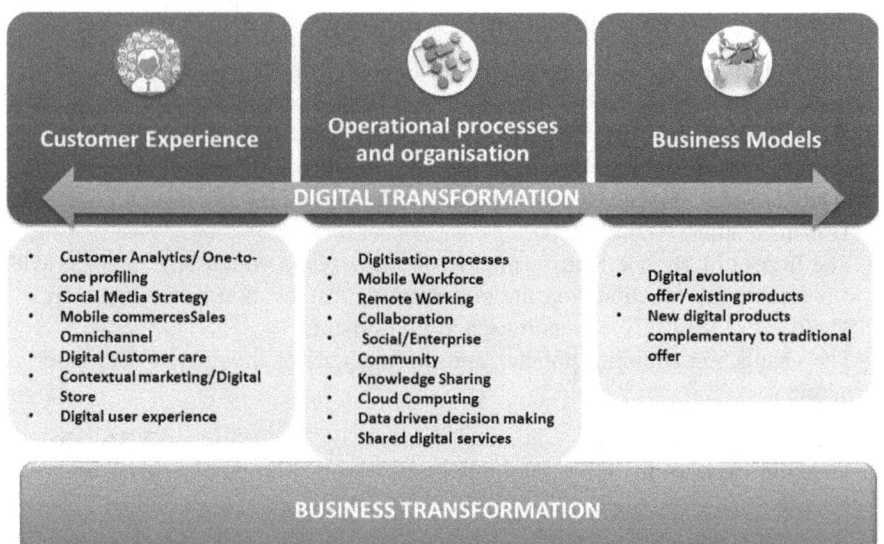

Fig. 2 Digital transformation is business transformation. *Source* NetConsulting cube

(c) The introduction onto the market of new and previously non-existent products and services conceived through the incorporation of digital technologies or the innovation of existing ones using this same method.

In the light of these assumptions it is clear that CIOs understand Digital Transformation not as mere technological innovation but as a structural transformation of the entire company in its organisational models, internal processes and market positioning.

4 Lesson 2

4.1 Digital Transformation Should Be Developed on the Basis of a Strategic Plan that Involves the Entire Company, from the CEO Down

The CIOs surveyed are convinced that Digital Transformation will have a major impact on the company at all levels, and that we need to accelerate its implementation because business is already being affected by the external environment (51%), and will be even more so in 2018 (25%) and 2020 (25%).

...a quarter of Top Managers have truly understood that we need to change in a digital logic. ICT and Marketing lead this change process.

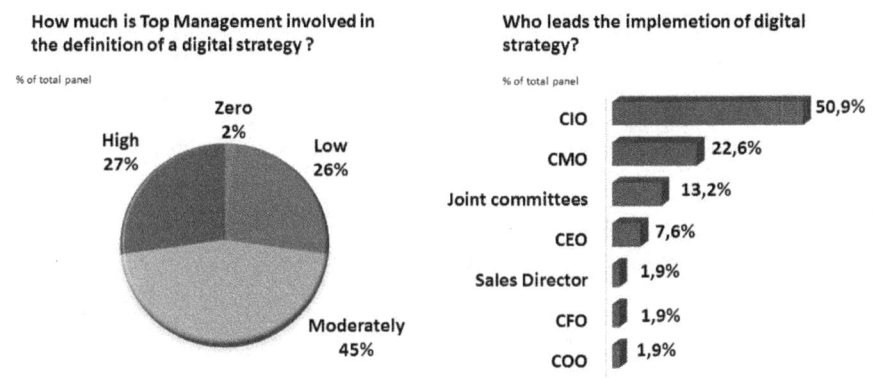

Fig. 3 Is top management committed to digital transformation? *Source* CIO Survey 2016

The strategic value that CIOs attribute to Digital Transformation, understood as Enterprise Transformation, and the urgency in its implementation, necessarily require the involvement of the CEO and top management.

However, the Survey revealed that only in 27% of cases is top management significantly involved in the implementation of Digital Transformation. It has little involvement in 26% of cases and is moderately involved in the other 45% (Fig. 3).

Underlying these answers, we can see that on the one hand top management sees Digital Transformation in predominantly technological terms and has little understanding of the business benefits it can bring, while on the other hand they fear that it will demand a very high volume of investment, the return on which will only be achievable in the medium to long term.

This is a cultural gap that CIOs, who lead Digital Strategy in 50.9% of cases, must help Top Management to bridge.

At present, the CIO, being the only figure in the company who fully understands business processes and technological infrastructure, may assume leadership of its digital transformation in the above-mentioned sense, also involving Top Management in this. But to lead Digital Transformation the CIO must constantly monitor the evolution of technologies and learn to sense the right time to incorporate them into plans for the innovation of the company's processes and evolutionary strategies.

To do this, the CIO needs to acquire transferable skills and soft skills, to spread digital culture in the company and build consensus and cross fertilisation relative to the Digital Transformation plan.

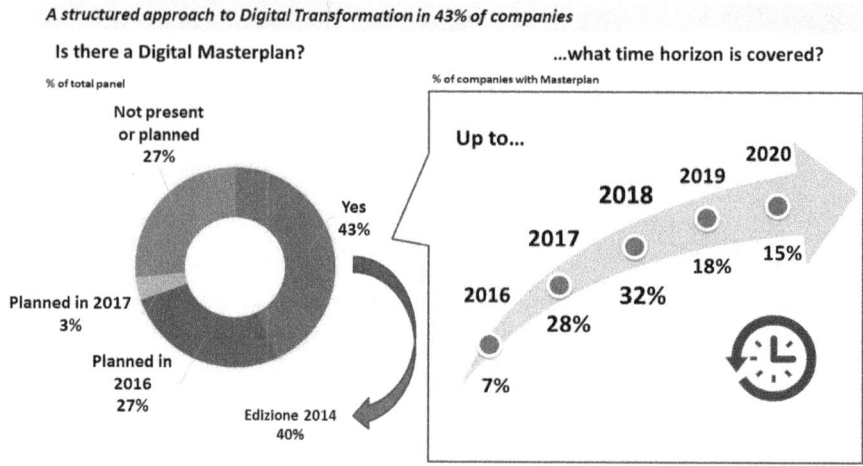

Fig. 4 Has a master plan for digital transformation been prepared? *Source* CIO Survey 2016

5 Lesson 3

5.1 The Path of Digital Transformation Should Be Guided and Directed by a Master Plan

The strategic nature and speed of implementation of Digital Transformation require the preparation of a master plan based on an "Enterprise Driven" systemic approach which, taking a strategic view of the company's innovative path in the medium term, clearly indicates the achievable benefits, the impacts on processes and the organisation, the investments to be made and the related economic returns, defining a roadmap of objectives to be met over time.

Forty-three percent of companies consciously developing Digital Transformation and participating in the survey, already have a master plan, while a further 27% intend to adopt one by 2016, and another 15% by 2020 (Fig. 4).

Only 27% of companies responding do not have a master plan.

6 Lesson 4

6.1 Digital Transformation Requires the Contribution of All Business Functions and the Lowering of Barriers Between CIOs and Business Managers

From an enterprise-wide perspective Digital Transformation must be achieved through the contribution of all main business functions in a climate of cooperation and mutual trust.

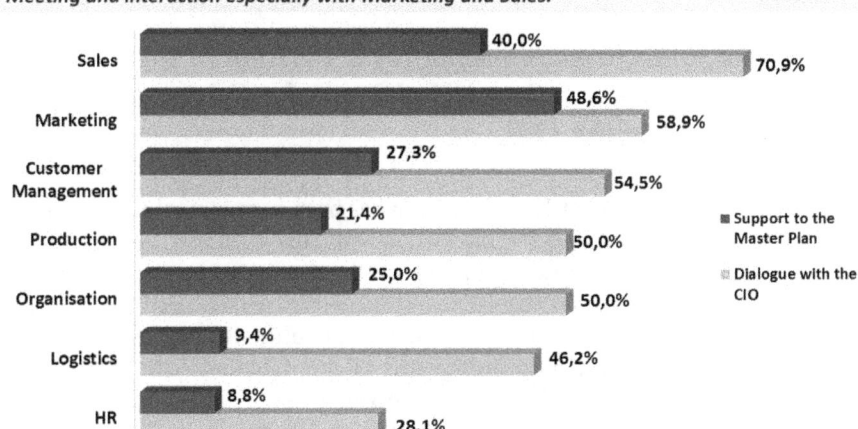

Fig. 5 Who does the CIO speak with most and who has contributed to defining the digital transformation master plan? *Source* CIO Survey 2016

To date, Marketing & Sales has developed more Digital Transformation projects than any other company function, guided in this by the need to interact with digital customers on the basis of the paradigm of customer experience.

This has occurred through often poorly coordinated initiatives between the CIO and Head of Marketing, not always in a climate of cooperation with the CIO.

This situation is rapidly changing and the CIO Survey 2016 showed that all major company functions, not only Marketing and Sales, provide support to the master plan and dialogue constantly with the CIO (Fig. 5).

The accelerated and pervasive way in which ICT is penetrating all major company areas on the one hand helps to involve the managers concerned in ICT investment decisions by giving them budget and spending autonomy, and on the other hand, precisely because of this, tends in some cases to result in a misalignment in the choice of solutions and suppliers with respect to the development plans defined by the CIO and company information system compliance and security policies. It is clear that this process of delegation of expenditure and technology choices to business functions requires harmonisation with the CIO's plans and a new governance model that refers to the implementation of the Digital master plan and the involvement of top management.

The construction of an ecosystem of internal cooperation between functions is essential for the development of digital transformation projects where technology and business are intersected natively.

This approach is also important for the construction of a climate of internal knowledge as regards the implementation of the solution and in its subsequent adoption.

7 Lesson 5

7.1 Digital Transformation Requires a Revision of the IT Division's Organisational Model and the Corporate Positioning of the CIO

While on the one hand Digital Transformation has been promoting the development of a technology infrastructure for interaction with the customer and the digital partner, on the other it requires a thorough modernisation of legacy systems and applications.

The bimodal approach accurately responds to these needs, considering aims of efficiency, security and reliability on the part of the traditional core of the company's information system and aims of agility and speed necessary not just for operations but also for real-time interaction with the digital customer.

Forty-one percent of the companies surveyed already need to manage bimodal IT and are planning a further adoption over the course of 2016 (Fig. 6).

The bimodal approach also requires a strong synergy between IT and business and the consequent reorganisation of the traditional organisational model based on a clear distinction of roles between demand and the CIO.

Thus, it is not IT that listens, gathers requirements and proposes design developments for approval, but it is the joint IT and Business team which, with its vision of the business strategies/objectives and the different forms of evolution driven by technology, analyses the context, makes appropriate assessments and takes the relevant decisions. Cross-functional teams are especially common in Finance and in the Telecoms-Media sector.

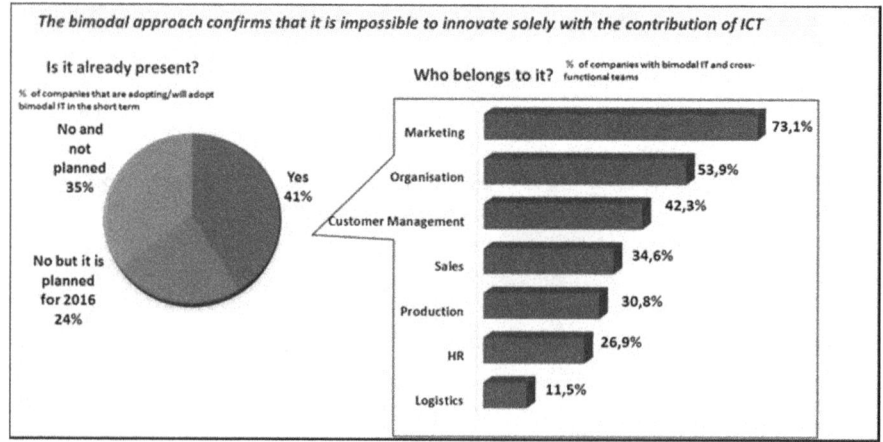

Fig. 6 Is there a cross-functional team to manage this evolution from a bimodal perspective? *Source* CIO Survey 2016

It is obvious that this way of operating mirrors the speed and agility that are typically at the base of a bimodal IT model.

The 41% of companies that are already taking a bimodal approach have formed cross-functional teams, with a particular presence of Marketing and Organisation managers, whose task is to carry out the native design of Digital Transformation projects.

8 Lesson 6

8.1 The IT Budget Must Become Dedicated to Digital Transformation Shared Between the CIO and Business Managers

In most of the companies surveyed the path towards Digital Transformation is in a phase of transition where while on the one hand, as previously stated, cross-functional teams have been created, on the other the IT budget is not shared equally but, on the contrary, many company functions have their own IT budgets, often allocated to digital projects and suppliers whose conception and whose choice is not shared with the CIO.

The CIO Survey 2016 shows that the IT budget is still mainly in the hands of the CIO, but that the share allocated to business functions increased by 100% between 2014 and 2015, with significant differences between the various economic sectors (Fig. 7).

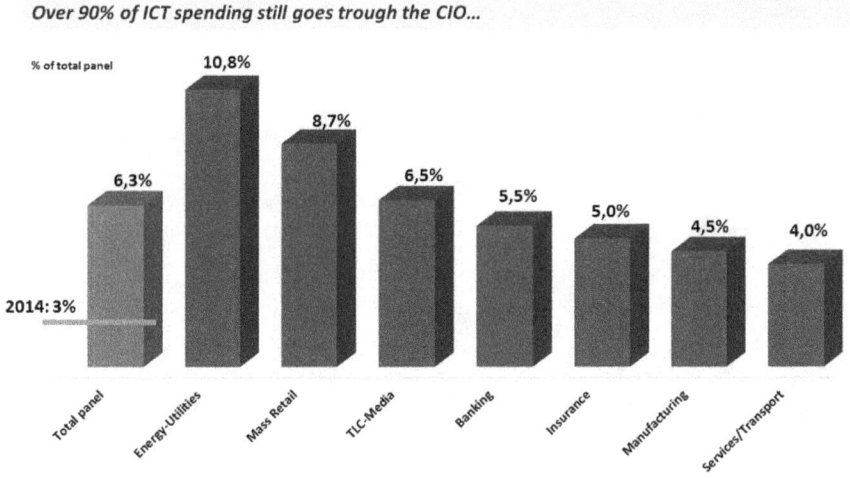

Fig. 7 What proportion of total expenditure is handled outside the ICT department? *Source* CIO Survey 2016

This is further evidence of the pervasiveness of ICT and the consequent redistribution of the budget available to business managers, with the previously described possible dysfunctional elements in their relationships with CIOs.

As previously mentioned, we need to create a consensual decision-making process as part of a new governance model for the innovation of processes, products and approaches to the market supported by digital technology.

Looking forward, in line with the establishment of cross-functional teams, and with the introduction of governance models for Digital Transformation, we need to define a budget shared between IT and business functions.

9 Lesson 7

9.1 Given Their Strategic Importance for Companies, Digital Transformation Projects Need to Be Evaluated in Terms of the Business Benefits Delivered, and not just the Technological Ones

The assumption that Digital Transformation projects bring structural changes to strategies and corporate performance has two types of consequences:

- That the return on investments benchmarks relate not only to IT but also to the gains made at a corporate level
- That, consequently, a cross-functional team in which the company's top managers are represented carries out the evaluation.

Companies are not too far away from this model in that although the Survey shows that in 48% of cases the CFO is the evaluator, and this is obvious because the first

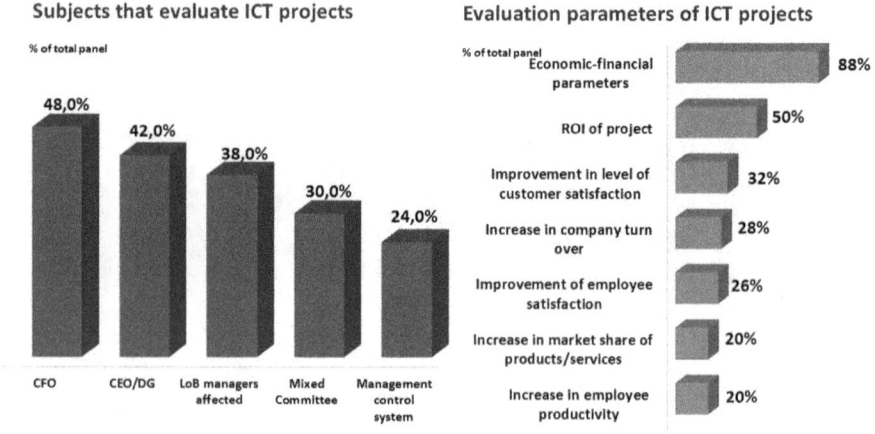

Fig. 8 Subjects that evaluate ICT projects and evaluation criteria. *Source* CIO Survey 2016

benchmark is compliance with the budget, it should be noted that the CEO or General Manager comes second (42%), followed by the Lines of Business (LOB) managers (38%), and that the most important benchmarks include "improving Customer Satisfaction" (32%), "increasing the company's turnover" (28%) and "improving internal Customer Satisfaction" (26%) (Fig. 8).

In general we can see that digital transformation, by enabling the company to design and sell new products and services on the market, as well as supporting the innovation of internal processes, reconfigures IT from a cost centre to a profit centre, structurally modifying its ROI indicators.

10 Lesson 8

10.1 Seeking Out Figures with New Skills and the Re-Skilling of Existing Ones Are the Most Important Enablers of Digital Transformation

It is commonly believed that the success of a Digital Transformation project depends largely on the availability of figures with new skills in the company who, on the one hand, have in-depth knowledge of the digital world and on the other, have the vision and ability to incorporate and make optimal use of technologies in specific company processes.

The ideal profile that emerges is of someone who combines technological skills and business expertise, which is very difficult to find on the labour market.

Consequently, separate skills are sought, the most in demand of which are Data Scientist, the IoT Specialist, Enterprise Architect but also the Business Analyst and Digital Media Specialist, to name a few.

Faced with the need for figures with new skills, who are hard to find and therefore particularly expensive and not easily accessible to small and medium-sized businesses, we need, especially in large organisations and public administration, to embark upon re-skilling projects for skills on their way to obsolescence or to reduce the number of surplus workers. The latter is more difficult in countries where the labour market is less flexible and this is an obstacle to Digital Transformation.

CIOs respond to these constraints either by training figures with new skills or by using the skills of the ICT Vendors, acquiring for themselves generic skills of a non-technological nature.

Moreover, the people-centric approach that is frequently adopted in Digital Transformation plans demands that the CIO possess an unprecedented ability to manage human resources in order to create sharing and agreement about the objectives and innovative projects.

11 Lesson 9

11.1 Digital Transformation Requires Support from External Partners from an Ecosystem and Open Innovation Perspective

Digital Transformation is a complex project that, involving both technological and business innovation, requires extensive and sufficient knowledge on both fronts.

At present, few companies have all the resources necessary for the independent management of such a complex process and so they increasingly rely on the support of external parties and position themselves within innovation ecosystems (universities, research centres, start-ups and innovative companies) within which they exchange ideas and start new projects.

The Survey showed that over a period of five years the profile of external actors perceived as capable of supporting important Digital Transformation projects changed significantly (Fig. 9).

While in 2010 the main contacts were ICT Global Vendors (75.2%) and Specialist Vendors (53.6%), in 2015 Specialist Vendors ranked first (73.6%), followed by Digital Agencies (55.3%), while ICT Global Vendors had dropped by around 25% points to third place in terms of mentions by respondents and ICT Consultancy Firms were also down about 10 points (from 39.4 to 29.2%).

Innovative Start-ups were the biggest most significant growth area according to the CIOs surveyed, gaining about 22% points from 15.7% of mentions in 2010 to 47.1% in 2015, moving into fourth place, ahead of ICT Strategic Consultancy Firms.

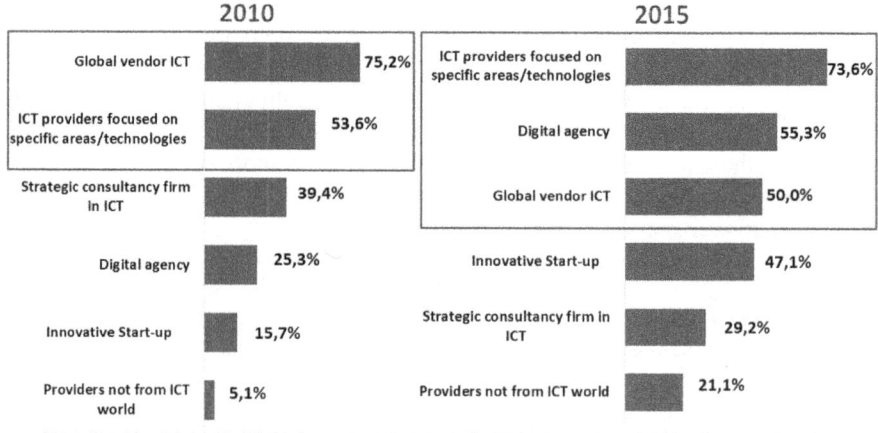

Fig. 9 Type of suppliers that support the company in innovation. *Source* CIO Survey 2010 and 2016

12 Lesson 10

12.1 The CIO and not the CDO (Chief Digital Officer) Should Assume the Leadership of Digital Innovation, Taking on a More Strategic Role

Italian companies participating in the CIO Survey panel stressed that obstacles to the Digital Transformation process are not economic in nature, due to limited budgets, but are mainly cultural, or rather a lack of understanding of the potential benefits.

This means that priority is given mainly to traditional maintenance projects and the evolution of existing digital tools (Fig. 10).

A reductive understanding of Digital Transformation in essentially marketing and sales terms is leading many companies to create the figure of Chief Digital Officer (CDO), separate from that of the CIO, often reporting to the General Manager or Chief Executive Officer Eller (2016).

This seems to be a partial answer to the technological and business complexity of Digital Transformation for several reasons (Fig. 11).

First of all, CIO and CDOs can possibly end up with overlapping roles and innovative parallel and potentially conflicting paths that make the Digital Transformation process slower, partially ineffective and difficult.

On the contrary, in this scenario, CIOs can and should take a leadership role in the digital innovation of the company in which they operate:

- Creating internal culture, above all at managerial and executive level
- Developing knowledge and skills about the business aspects of the company

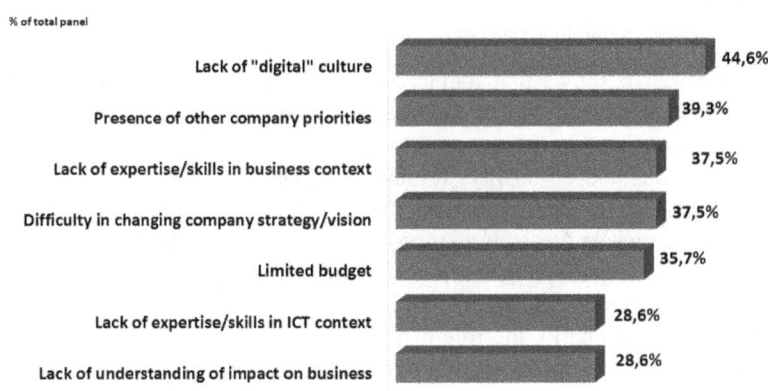

Fig. 10 What limitations does the CIO need to overcome to direct digital transformation? *Source* CIO Survey 2016

The Chief Digital Officer: not widespread and where existent still in search of their true identity.

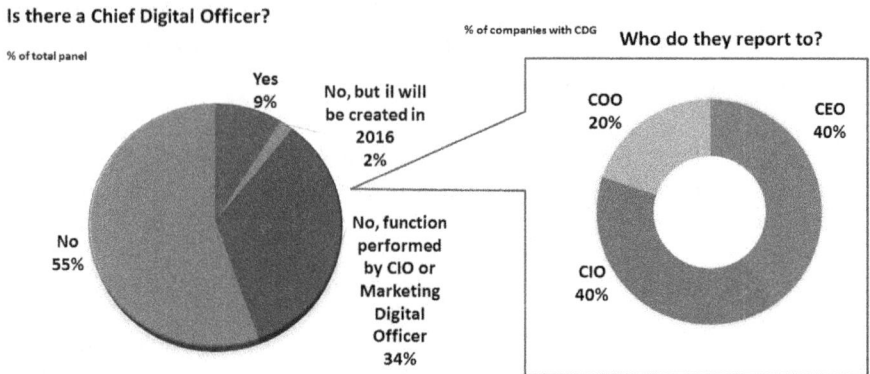

Fig. 11 Is there a chief digital officer? Who does this figure report to? *Source* CIO Survey 2016

- Establishing and generating cross-functional teams
- Initiating a permanent observatory on technological developments and success stories
- Initiating relationships with innovative parties and setting up their own network or ecosystem of innovators, beginning with the start-ups
- Participating in Strategic Committees.

By building on these lessons learned the CIO community could make a significant contribution not only to the Digital Transformation of their own companies but also to that of the entire national economy.

13 Conclusions

The CIO Survey 2016 shows that Italian companies of all sizes are now moving from a general interest in digital technologies and their potential impact on the business to the launch of projects—projects that are not always clearly aimed at structural innovations and not always framed in a master plan.

The Digital Journey remains a process that is not yet clearly structured, whose expected results cannot be precisely mapped in advance, which leads companies to adopt a prudent and experimental attitude towards major investments in "disruptive" Digital Transformation projects.

In this transitional scenario, the role of the CIO can be positioned between two opposite extremes within the company, that of manager or weak innovator of existing technology infrastructure or that of the Digital Transformation leader whose task is to innovate the business through the intelligent use of new technologies.

However, this latter aim can be achieved under two conditions: that the CIO has the courage to question the constraints of budget and prospective vision that typically limit their willingness to innovate, and that top management and CEOs understand that digital transformation is the only way for companies to regain competitiveness and sustainable growth.

References

NetConsulting Cube (2016), CIO Survey 2016
Uhl A (2014) Digital enterprise transformation. Routledge
Eller M (2016) Be the business: CIOs in the new era of IT. Routledge

CIOs at the Centre of a New Humanism

Bruno Demuru and Teodoro Katinis

Abstract This chapter addresses an evolution in corporate organisations that was unthinkable until recently and that represents an important opportunity for changing philosophy and practice in business. We will focus principally on the CIO (Chief Information Officer), who is at the centre of this evolution, and on the characteristics of his/her new central role not only in IT but in all aspects of business. The historical approach at the beginning of the article aims to summarize the key events and turning points in the evolution of the business organisation, which will serve as preparation for the theoretical approach of the second part. Humanistic discourse intersects with and contributes to the development of both the historical and the theoretical approach presented in this chapter. Specifically, the chapter addresses the tendency to simplify the complexity of a real world, a need that should not be completely condemned but rather reassessed for the advantages and disadvantages it brings to the corporation's organisation. The simplification process tends to marginalize human beings and their complexity, while the new approach we propose aims to put people at the centre of the process. Choosing between complexity/complication and simplification means opting for either a traditional or an innovative approach to business and the role of IT, led by the CIO, in a company. We will analyse the impact this re-ordering may have on a company, particularly on productivity and profitability. Furthermore, we will try to understand the implications and consequences of the recent technological evolution and how to benefit

---

The views expressed in the paper are those of the author and do not necessarily reflect those of the company.

B. Demuru (✉)
Saras Group, Milan, Italy
e-mail: bruno.demuru@saras.it

T. Katinis
Ghent University, Ghent, Belgium
e-mail: teodoro.katinis@ugent.be

from it. Among other topics, this chapter will mention the practice of collaboration among workers, the use of big data, and the "Bimodal" approach in light of new humanism. Looking at the future of companies, the authors suggest the evolution of the scientific organisation into a humanistic organisation, where new figures should guide this exciting transformation.

1 Introduction

The evolution of digital technologies presents new challenges and opportunities for organisations. On the one hand, the speed and magnitude of recent technological developments are so unprecedented that they call the well-established foundations of corporate management into question and coerce organisations to change. On the other hand, new technologies provide a needed stimulus for organisations to transform their business and exploit growth opportunities. In this transformational process, the central role is assigned to the company's CIO (Chief Information Officer) who is expected to drive and implement change within an organisation. Prior to initiating the process, however, each CIO needs to decide which "grand" principles he/she intends to follow, as these principles will determine his/her priorities and guide his/her choices throughout the transformation.

In this chapter, we discuss and elaborate on a new philosophical concept that has been recently introduced in business and IT press: digital humanism (DH). As opposed to digital "machinism"—a perspective that regards the automation of human work as a primary focus of technology—digital humanism emphasizes the role of people in technology and views technology as a means for solving human problems and addressing human needs. According to Gartner's DH Manifesto, the new organisation of business should "start and end with people" and "embrace serendipity", which suggests that the new CIO, as a central figure in the re-humanization of a company, will embrace and guide the process toward a DH approach.

To avoid confusion, we would like to draw a distinction between the notions of digital humanism and digital humanities. Scholars in the field of digital humanities integrate digital technology into the humanities disciplines by employing digital tools to produce, store and access knowledge, for example: (1) to conserve texts in digital copies, so that one can recover them even when manuscripts or books are absent or corrupted; (2) to consult texts from different places and by different devices without needing to possess a paper copy; (3) to search within texts for specific elements for research and/or didactic purposes. In this chapter, we would like to propose what we believe is a much more revolutionary way to conceive the relationship between ICT and humanities by proposing a new definition of the CIO's role at the centre of a new humanism.

2 Digital Humanism Between Scholarship and Industry

Before examining the CIO's transformation into a modern humanist, we need to clarify the meaning of expressions such as "humanist" and "humanism". In a broad and more general sense, "humanism" is an approach to the world that emphasizes the importance of a human as an object of theoretical and practical inquiry. In a specific sense, "humanism" refers to a long-lasting and heterogeneous early-modern cultural revolution that began with the 14th century humanist Francesco Petrarca and peaked in the Renaissance period with Italy as the first and most important centre of diffusion in Europe. "Studia humanitatis" is the central concept for understanding the cultural revolution that took place in early-modern Europe. With that expression, pre-modern scholars meant to put mankind, in all its aspects, at the centre of their work, from theory to practice, from philosophy and history to politics and ethics, from poetry and rhetoric to science and technology. This interest was expressed in different ways with a common starting point: the rebirth of the study of ancient classical works in all fields as the basis for reinterpreting and changing the contemporary world; in other words, the rigorous study of the past put at the service of the present.

We believe that the revolution of the CIO role we are proposing in this chapter shares some similarities with both meanings of "humanist"/"humanism" mentioned above. Indeed, as a new humanist, the CIO exhibits a concern for the people and human resources involved in the processes under his/her control and appreciates the necessary tight connection between the study of cultural heritage and the present world for promoting a revolutionary approach to work and business.

3 The Scientific Organisation and Its Historical Evolution

Historically, early industrial organisations were predominantly concerned with optimizing work processes to minimize inefficiencies, eliminate wasted effort and, consequently, to increase the profits. Little attention has been paid to the role that human resources might have in these processes.

Business organisations of the twentieth century evolved according to a different, scientific kind of logic. Already from the end of the nineteenth century, the theories of Frederick Winslow Taylor, followed by the practical applications by Henry Ford, introduced the concept of scientific management, which in those days was still mainly aimed at maximizing productivity. In this period, the power of the United States of America started to burgeon, and the basis for their surpassing the European countries in this respect was laid. Briefly but emphatically, the weak points of this evolution were manifested dramatically in Wall Street's 1929 crisis, and its catastrophic consequences in the subsequent years. Nevertheless, the growth in social well-being was enormous and, thanks to these successes, the dissemination of a scientific mindset through corporate organisations started to shape our world and spread from West to East.

In the early '60s, the first computers (IBM, UNIVAC, and others) started to play a revolutionary role in corporate organisation. These early years of cybernetics, intelligent machines, expert systems, along with a cultural and philosophical vision of corporate reality led to a preference for purely logical and rational thought that did not take into consideration the more humanistic characteristics involved in the organisation.

A possible early critique of this business re-ordering could be found in the "first generation" of scholars from the School of Frankfurt for Social Studies, in particular Horkheimer and Adorno, authors of the popular Dialektik der Aufklärung [Dialectic of Enlightenment], published in 1947. Since the '30s, Horkheimer, the founder of the so called "critical theory" and father of the School, had been promoting an analysis of the contradictory nature of the Enlightenment project and its ideal of an absolute sovereignty of reason. According to his theory, during the eighteenth century, and in particular with the philosopher Immanuel Kant, reason became a new religion that brought the Western world towards the tyranny of scientific thought and technology. The tragic consequences and the human costs brought by the modern tendency to worship the god Reason, such as the massacres of the French revolution and the World Wars, were in front of everyone's eyes.

Fordism, the post-WWII industrial paradigm initiated by the American industrialist Henry Ford, is characterized by the mass production of standardized goods produced on a moving assembly line. This system introduced both mass production and mass consumption. Of the company organisations that applied Fordism for to production, the most popular is, perhaps, McDonald's. Praised by some and condemned by others, Fordism may be considered a main target of the School of Frankfurt's attack.

For as long as organisations continue to follow a Fordism approach towards organising and rationalizing the role of IT within an organisation, the critique of such a mechanistic approach will still apply as it did in the early '60s. We believe, however, that today's digital world provides massive opportunities for "humanization" of IT and that the CIO can be regarded as a new humanist of the 21st century.

Both medium and large size businesses have made repeated attempts to stage a qualitative leap by introducing new concepts linked to humanism and the central role of human resources. It is not uncommon to hear human resources being referred to as "true corporate stakeholders" of a company, partly in an attempt to keep up the productivity and increase the motivation of the company's employees. It should be said, however, that such endeavours, as praise-worthy as they are, have remained episodic and have never really become firmly established within the company culture.

4 The Concept of Simplification in Organisations and Its Critical Points

The scientific organisation that was born at the end of the 19th century is based on certain conceptualizations that are generally used for the planning and creation of corporate organisations. They are based on a simplified representation of a company as a complex system and regard the employees as parts of a machine that perform specific tasks and jointly contribute to the functioning of a complex whole. As such, these conceptualizations seem to suggest that human beings, in all their complexity, are not at the centre of the company. While in some cases this perception is correct, in other instances, people are, in fact, the "protagonists" in a company. In such cases, using a machine mechanism analogy to represent a company may be misleading and too simplistic, which brings us to a reflection on the way we represent the corporate organisation that affects our understanding of what the company actually is or should be.

A company's organisational structure is usually pictured as a hierarchy with different levels from top manager down to staff and employees. These hierarchical structures may get rather complex; even small companies may encompass 7 or 8 internal hierarchical levels. At the same time, we tend to represent corporate processes as running like clockwork. These processes include the interaction between different roles and functions, the flow and integration of activities, the interactions between various personae (according to the configuration of their corporate roles), and the way in which people in a specific role start an activity that must subsequently be completed by another corporate role. Indeed, the metaphor of a machine, or a chain of production—in the Fordist fashion—represents the functioning of an organisation composed of actual people.

If the wheels are functioning properly, our clock works precisely and perfectly. Still, we should ask ourselves if people are really comparable to the cogs of a clock, or parts in a machine. If we assume that the best representation of a company and its people is in fact a machine, even a very complex one, we risk missing all the humanistic aspects of the company and excluding the most important protagonists in the organisation: the actual people and human resources.

First, a linguistic issue needs to be addressed before any further analysis of the new type of organisation we want to support. We should pay more attention to the figures of speech we use to communicate the image of the company within its boundaries and to the external environment. The rhetorical aspects of communication affect the way in which we represent a company and its organisation. The "clockwork" we mentioned above is a clear example of the relevance of rhetoric in this regard. Let us assume that a revolutionary approach to business requires a revolution in communication. If that is the case, we should avoid describing the new ideal business organisation in outdated language (i.e., the language we have used until recently). In other words, the change we are proposing is accomplished, in part, by modifications on a rhetorical level. If the way in which we represent the world matters and affects how we perceive it, then we need a new business language for a new dimension of business. Historically speaking, the metaphor of the clockwork—to remain consistent with the figure of speech often used in several academic and industrial fields—was used in the Western world by modern rationalists to describe a universe built by God as a perfect mechanism. Against this representation stands another one proposed by Renaissance philosophers and humanists and, after the age of Enlightenment, in the Romanticism of the eighteenth and nineteenth century. Their non-mechanistic representation proposes using an organism, and its living dynamics, as the most appropriate metaphor to describe the world. Within the history of Western civilisation, these two opposing models have been the most important representations of the universe and its "parts" (or "organs," it depends on which model we want to use).

We would not be able to understand the nature and history of the mechanistic and non-mechanistic model without employing such metaphors. As the German philosopher Friedrich Nietzsche argued, all language stems from a process of metaphorization of reality, from perception to intellectual conception and linguistic codification. Put simply, the way we speak affects our knowledge, understanding and judgement, and how we share them in a specific environment.

If figures of speech matter so much in our human experience in general, then they will matter equally for business organisations. To that end, we should recognize the power and importance of language and start using contemporary rhetoric when trying to convey our business vision of the future—a human-centric business with the CIO as a coordinator at different levels.

A mechanistic representation of corporate reality also affects our approach to education and training. They way business disciplines are being taught—in the classroom and online—emphasizes the importance of studying a well-defined set of procedures and rules and learning how to apply them. Furthermore, the system of control and evaluation follows a similar rigid approach: failure to adhere to the rules is sanctioned and most failures are viewed as consequences of not following the prescribed procedures. Success is thus measured in terms of how strictly one can follow the rules. When we change perspectives, however, we realize that it is extremely difficult to follow an onerous and complicated body of rules.

Along with this body of procedures, other scientific tools are also used in organisations, such as safety plans, organisational models, and a control matrix. In Italy, the legislative decree DLS 231/01 represents the main standard for safeguarding the civil and criminal liabilities of companies, and it helps to demonstrate compliance with certain requirements. The international standard ISO 9001, to broach another subject, represents the "best practices" for working at a quality level, as it defines rules for managing the functioning and production of final products, endowing them with the best possible characteristics as far as quality is concerned.

Applying these internal standards is extremely complicated. We often hear from staff that quality is a necessary aspect, something that must be accepted, but at great cost. Either the current regulations or the market requires quality standards, and so quality is perceived more as a burden than as a solution to problems.

One might argue that the guiding principle behind introducing these rules and requirements is the human need to simplify the real-world complexity. Simplifying the complex reality affects our understanding of its important aspects, especially when we deal with "systems" in which the human resources play a major role, like in a company of any type.

Following only the rational approach, we risk underestimating the role of a human factor that cannot be modelled by a mathematical algorithm. Simplifying helps with understanding some aspects of the organisation, but we should always keep in mind that this is always a partial representation that cannot be confused with the comprehensive reality of an organisation.

As we mention at the beginning of the chapter, the School of Frankfurt pointed out the moral and social costs of modernity and the modern approach to living. The authors of the School focused on the macro events, such as the Worlds Wars, but not on human costs in the work environment. The very beginning of Dialectic of Enlightenment (1947) pictures the relationship between the Enlightenment and its tragic consequences in a few efficacious sentences:

"Enlightenment, understood in the widest sense as the advance of thought, has always aimed at liberating human beings from fear and installing them as masters. Yet the wholly enlightened earth is radiant with triumphant calamity" (p. 1).

In other words, according to the authors of the Dialectic, although the Enlightenment was conceived as a new humanism to give human beings the power to freely determine their present and future, the actual results were a de-humanisation of the world. The destructive power of science and technology was, according to the text, the most relevant aspect of modern thought and approach to reality. We may apply this critique to the development of business organisation so far, highlighting its lack of humanism due to an excessive use of the mechanistic model. Nevertheless, we believe that, nowadays, we have the power to invert the tendency of modern organisations, based on a mechanical approach to reality, and put IT, which is at the peak of the technological evolution, at the service of a new humanism.

5 The Impact of New Technology

Over the last few years, we have witnessed the global phenomenon of a technological revolution, which is creating a real upheaval in the traditional ways of doing business. In this regard, one must reconsider the existing organisational structures of companies that, until now, have been based on the scientific and oversimplified approach we described briefly in the previous paragraphs.

We defined this phenomenon "technological revolution" because it has rapidly made new and powerful technological tools available for organisations. The exponential increase in the availability speed and consequent supply of these technologies has forced companies to question the consolidated paradigms, and to push for an organisational redesign. It has already been a while that, at an international level, some important concepts have been introduced, such as digital disruption, digital business transformation, industry 4.0, the status quo challenge, and many others. These concepts touch upon the function of the organisation that was intimately tied to digital technology and that had been traditionally appointed the role of harbinger of propositions for a continuous stream of new digital tools: this function was, and still is, ICT (or, to give it its updated name, IT).

The arrival of the new technologies and their implementation in business demand an organisational and cultural redesign of the company, particularly of its IT. Before we describe this change, we'll try to summarize some of the principal technologies available on the market now, that are evolving with the extraordinary speed mentioned above.

5.1 Big Data and Analytics

Big data, as reported by Wikipedia, is the term used to describe a collection of data that is so vast in volume, speed, and variety that you need specific analytical technology to store, process and analyse the data and then extract valuable insights. These technologies are available today and are increasingly being developed with impressive speed. These new technologies are more effective than previous tools for analysing the large amount of data that has become available. In the paragraphs below, we will make particular reference to corporate organisational technologies.

The tools for discovering, interpreting and communicating an analogous and meaningful model of a given system are subsumed under the term analytics. While the analytical approach gives us tools that were unthinkable before, true innovation only happens in the event of a veritable cultural change in human habits and behaviours. The lack of such a change explains the real obstacle to the large-scale application of these technologies today.

In the United States, the use of Big Data was launched some years ago, through various applications, in different corporate and political processes; while in Italy, for example, it faces a lack of cultural acceptance.

The greatest problem is that big data analytics are being treated as the exclusive territory of Information Technicians, largely disconnected from other business functions. The results obtained through the rigorous analysis of big data are often discarded and substituted with conclusions that are based on intuition, heuristics and common sense—modalities that top management has been using for decades in their daily working practice. These old and traditional decision-making modalities are so deeply rooted in the company that they might stand in the way of a potentially cutting-edge innovation within a Company.

5.2 Knowledge Management

Knowledge and its management within companies have been among the primary concerns of traditional and scientific organisations for years. What does knowledge mean for traditional companies? It means creating order in the flood of documents, experiments, developed technologies, patents, competencies, and so forth, which a company elaborates for its daily business. But knowledge in a more humanistic sense goes beyond organising the library and optimizing the document flow. According to the humanistic learning perspective, people's access to knowledge is not exclusively focused on pure ordering and categorizing but also on the meaning of words and texts. This means entering in the field of semantic search engines that overcome the limits of statistical search engines (with Google on pole), as powerful as the latter may be. The underlying logic behind a semantic search engine and the competences required to build such an engine would differ substantially from those required for designing a statistical search engine, and are much closer, yet again, to humanistic approach.

5.3 Collaboration

Social, or collaborative, tools contribute substantially to building and developing the culture of collaboration within companies. Collaboration has been traditionally underestimated and often ignored in corporate projects of a scientific nature. Only a few companies, particularly those competing in the high-velocity markets that require high levels of innovation and creativity, have put these tools to use within their businesses.

In this context, we must distinguish collaborative tools for internal use from those whose purpose we would define as external, such as customer management (CRM or similar). Social tools help companies to manage the "real" organisations whose scientific architecture relies mainly on organisational charts, processes, and roles. These social tools represent the "hidden" fabric of corporate functioning, and surely its most humanistic component. They involve interpersonal relations among people, personnel's perceptions of colleagues, and the application of competencies that do not feature on organisational charts and job descriptions, as they are external to the official roles but still an asset of the company.

Take an expert, for example, who is transferred to a different role for organisational reasons; he can share, through these social tools, his experience and know-how with people who will benefit from his former role, even if they don't attain the level of competency and knowledge of the expert.

Collaborative tools allow one to face and understand, in teams, critical working areas. Many companies today use the "ethical code" to sum up corporate values and tools of collaboration. This code helps to measure the application of a merely theoretical ethical code, to an actual corporate system and the real, humanistic world of the company.

The corporate climate, the pro-active and innovative potential of staff, the capacity to delegate and control by management, and other aspects within a company can be monitored with these social or collaborative tools. This monitoring ideally enables us to analyse these aspects, understand them, and then proceed with activities that aim to improve performance.

5.4 Digital Work Place

The concept of Digital Work Place—a workplace that uses digital tools—has always been underestimated or oversimplified in the tradition of scientific organisations. This oversimplification is the result of the low level of importance attributed to the logic of information and knowledge, especially compared to the logic of finding practical solutions to problems and the need for decisional synthesis at the top levels of the hierarchy. Minimalizing its importance has been done to the detriment of a system that enables the autonomous creation of operative solutions in daily work, which can be done more effectively by people who possess an intimate knowledge of operative details. Consequently, decision-making has always been delegated to people who were well-aware of the operational details and can therefore analyse the operational processes, critically assess them and propose suggestions for their improvement.

5.5 Hybrid Risk Management

The methodology of Risk Assessment comes from the common framework founded on "best practices" (ISO 270001 in the field of computer security, for example, ISO 9001 in the field of the quality of productive processes, etc.) that have historically been advantageous to scientific organisation. The methodology deals with issues that today are becoming increasingly critical in the functioning of companies; having an effective way to deal with these issues is an asset of an increasingly digitalized corporate architecture, which necessarily requires things to be carefully tested for their configuration.

Management according to the "best practices" method allows you to control the system that is the object of the analysis. These tools, however, generally tend to overlook the immaterial phenomena on which the real operative functioning is based, and therefore the impact of human resources in handling the organisational processes.

The hybrid approach, applied already in Operational and Commodity Risk Management, is not currently widely used, except in the evolutionary phase of some research projects. Therefore, industrial practice is still quite limited.

The hybrid approach combines qualitative and quantitative themes for risk management, composing an integrated vision that comes closer to running an organisation and takes considerations related to human resources into account.

5.6 Consumerization

The phenomenon of consumerization has completely overturned the technological planning of information systems. In the early '90s, only a handful of companies had a computer available to manage their activities. Those that were better organised had an IBM or UNIVAC Mainframe, while mid-sized companies had more affordable and modest solutions. With the arrival of the PC, corporate information technology evolved towards client/server networks and the creation of personal computers for carrying out one's own calculations and responsibilities.

A basic issue that remained, though, was where to separate the activities carried out at the Company from those undertaken in the context of one's private life. The fundamental, non-humanistic, theme consisted of the assumption that there was a profound difference between the corporate and the private personae, and the idea that the two had distinct behaviours and tools permeated traditional logic.

Today, new technologies have provoked an outburst of consumerization. Smartphones—once an exclusive asset of activities carried out in private life—have been brought inside corporate life, by social network tools, tablets, free Internet access, and similar things. All these tools were previously considered part and parcel of one's private life. One can imagine the profound disappointment and worries of people who are more closely linked to the traditional cultural canons. In a way that was unconditioned by design, digital technology has broken down the barrier between corporate and private, pushing forward the concepts of BYOD (Bring Your Own Device) or the reconsideration of themes such as Privacy, a humanistic theme that often is merely "tolerated" by the corporate "scientific" models.

Today, the reinforcement of information-based culture within companies is also connected to the proliferation of information technology that, before, was limited to the private environment; this reinforcement carries with it important benefits for companies in the fields of the security, privacy, and use of passwords.

5.7 Digitization and Simplification of Corporate Processes

The Digitization and Automation of corporate processes have two goals in a modern Enterprise:

- Digitalizing information previously stored on analog mediums (e.g., hard copy) or, in the best of cases, in excel sheets, word files, or PowerPoint presentations.
- Pushing the system towards the automation of processes, thusly minimizing manual input and routine work.

The second point is connected to the concepts of work in the context of processes, and to the theme of collaboration described previously.

Once manual and operational workload (for example back office activities) is reduced, time and resources are freed up for performing other, more value-adding activities. Such activities may include advancing one's knowledge within one's area of expertise, learning new skills and competences, or searching for alternative solutions and opportunities for improvement of the existing processes.

Digitization projects have a fundamental prerequisite, however: the ease of implementation. Ultimately, the success of such projects is contingent on corporate processes being simplified and on the company becoming agile and "brisk". In short, it is essential to break down the walls of bureaucracy, one of the most resistant barriers in business, as it is born out of the fear of managing a complex system.

For clarity of argument, we must avoid confusing the concept of simplification, a fundamental asset of the scientific organisational approach described above, with the concept of agility and structural simplification of the organisation, a fundamental asset of humanistic organisations.

Today, many experts of organisations view bureaucracy as an obstacle to improving corporate performance. Still, many companies struggle to overcome this construct and continue to maintain a highly bureaucratized architecture: a less humanistic situation is hard to imagine. Even today, in 2016, many businesses suffer from this problem.

6 Restructuring IT's Function and Its Role in the Humanistic Redesign of Companies

The function of IT in a Company has a particular characteristic that distinguishes it from others: it typically supports processes. Because IT projects are rarely confined to a single corporate function, the job of IT is viewed in a wider context. The most modern IT organisations have abandoned the traditional role of administrator of applications, and have substituted it with the administrator of business processes. This distinguishing characteristic allows IT to remain above the level of single viewpoints when working in a company that has decided at its highest levels to truly change its operative modes and transform through the pursuit of actual and profound performance improvement. This is even more true if the company, in conformity with the development of its market of reference, decides to adjust its business model in order to obtain, within a limited time frame, actual results through its new operational modality.

According to the pattern of processes, to coordinate operations with business, IT requires a work model transformation from one of a technical support to one that is more business-oriented. The function of IT must evolve from its departmental design into a team of multi-skilled professionals that are both specialized in technology and competent in business. In order to make this happen, IT professionals must work to enhance their competencies. Doing so, however, will prove difficult, given that the IT team will have to continuously build new skills to remain updated on new technologies that are emerging on the market.

These new technologies must be used, in projects that reflect a new humanistic dimension in company operations, and that express, through their realization, a few fundamental rules that are listed below:

- Introduction of a vision that integrates the concept of user experience.
- Redesign of traditional interfaces for applications towards users.
- Adoption of operational models that involve users.
- Monitoring of user satisfaction and carrying out proper actions for improvement accordingly.

The IT organisation and management of new projects must be redefined according to these criteria. As always, in order to obtain concrete results within a reasonable time frame, we must look for a valid compromise between past and present to facilitate a transformation towards the future. In fact, we would like to present the IT transformation project in a way the American consultancy firm Gartner has called "Bimodal".

To proceed in this direction, companies not only need a new a type of CIO but also an up-to-date skill set for the IT personnel. The IT staff needs to develop know-how in entirely new areas: psychology, conflict management, change management, empathy, leadership, and the ability to sell oneself and one's solutions (i.e., the marketing of self). Furthermore, today's IT staff is expected to master the ability to manage client relationships both internally and externally.

We have discussed the transformation of IT architecture. We would now like to take a holistic perspective on the organisational framework. In doing so, we aim to answer the following question: when a company evolves from the old, scientific paradigm into a new, humanistic paradigm, should this transformation happen in a "destructive" mode vis-a-vis the past, or should it be conceived in a "lighter" and less radical way?

7 The Organisational Transformation of the Company

The company, or rather: the organisation of the Company "can" and "must" become humanistic. This imperative call is inextricably tied to the technological revolution that is underway. Whereas, in the past, the scientific approach could be justified by the necessity to simplify the complexity of corporate life, today we can no longer accept this argument. Also today, companies that implement transformational projects of such epic proportions need to understand what the main economical drivers of the activity are. On the one hand, the approach of a traditional "business case" is no longer so easily applied. On the other hand, traditional legacy architecture that was created and nourished over decades of a company's existence, cannot be radically substituted with a new one, but needs to be first paired with an alternative innovative approach that eventually will replace the old approach.

Indeed, even if IT, perhaps on the initiative of a visionary manager, tries to evolve in humanistic terms by itself, while the entire system continues to operate in a traditional mode, the project is destined to fail. An operation for the organisational and cultural redesign must be carried out throughout the entire organisation and with the support of top corporate leadership.

Also in this regard, collaboration plays an essential role in this transformation process. In our experience, not only have collaborative projects been appreciated by those involved, but they have also resulted in the development and practical implementation of a number of interesting operational improvements. By its own nature, collaborative activities promote initiative that fosters even more collaboration, which is a powerful booster for innovation in a company.

As mentioned before, the application of social media technology allows for the redesign of a traditional scientific organisation, on the basis of real data, in a humanistic way; this means that the projects would be highly connected to people's behaviour. In order to manage these IT projects, one must seek the added value in new competencies.

If we think back to the theme of consumerization, the idea is to spur the evolution of interfaces used in a work context towards a modality that is similar to home computers, that is, more "human" interfaces, that take into account the human factor in their design.

In short, the humanistic component of the "unexpected," of the "hidden," of the "irrational" needs to play a greater role in the reorganisation of the company.

But how is the head of the company and the leader of the transformation, the CIO, supposed to manage this change, supposing that he absolutely co-opts and supports it? This is the complex issue we are going to discuss in the next section.

8 The Evolution of the Characteristics of the CIO

From the perspective of the IT organisation transformation in a company, CIOs need to change by developing new competences and a more business-oriented attitude. This idea is not entirely new. Ever since the early '90s, at nearly all seminars related to information systems, it was reiterated that the CIO needs to change. More than 25 years have passed, but little has actually changed. Now the transformation is becoming a necessity because the consequence of not changing is quite clear: either the disappearance of the role of the CIO or its enclosure in a typically technological capacity, the role he held in the past. The arrival of new technologies and of the phenomenon of consumerization have created a situation in which corporate information skills have proliferated all sectors and have well overstepped the boundaries of the functions of IT, and particularly of the CIO.

If the CIO maintains his purely technological role, it would be better to re-dimension it to a simple CTO (Chief Technology Officer). It is easy nowadays to find CTO-services and the running of hardware infrastructure and applications on the market place. Many, even large, companies offer these services. Cost-reduction has become, more than before, one of the most widely observed modalities by scientific organisations to deal with the financial crisis.

Delegating the management and execution of technological activities to external companies is one of the main risks for a company today, especially when it comes to its core business. Losing control and governance of the information systems means losing control of the development of business itself, risking deadlock. Corporate business is increasingly becoming digital. For this reason, by pushing critical IT functions into outsourcing or, alternatively, under the hierarchical guidance of a function in a scientific-type organisation, one can obtain disappointing results and eventually hurt the corporate business.

In order to effectively govern the evolution that we suggest pursuing, the technical competencies of the CIO should be complemented by business knowledge and so-called "comprehensive interaction". That is, the CIO should learn how to listen to the requirements of users, and understand their needs and wishes that often are unconscious and therefore non-rational; he/she must increasingly become a psychologist. He/she must lead the company's push towards change, coherently and methodically, and have an evolutionary vision, inspiring respect and esteem in his interlocutors.

Furthermore, The CIO must learn to face complex changes with the appropriate serenity and the right methods. Therefore, he/she must be persuasive (like the ancient sophists) in applying rhetorical skills to convince business users, gathering their consensus to accomplish a common evolution for their own benefit. But to do so, the CIO must be ready to face new insight, accept the culture of analysis and understand other people's language and perspectives: open-minded, indeed, and open to accepting the challenge of a changeable, complex, and pluralistic human environment.

One of the most important themes that involve the CIO today is the so-called Demand Management, i.e., the management of the demands of internal users. The internal users traditionally submit their requests for solutions to him with an attitude that leaves little space for real change. They simply desire to have some new technologies, perhaps because they are modern, or in fashion. In the past, when IT was weak, this has contributed to the creation of so-called application "legacies," or the modification of systems like SAP, according to the model "customizing," which consists of bending applications that were built in a standard mode towards the working modalities present in an organisation. The information system gets applied, but matters were not truly changed.

The challenge for the new CIO is to gain a perspective of "governance" over the development of information systems, but not as he did in the past. To do so, he must channel the demands of business towards a logic of change that is to include the process itself, and not merely its tools. The overall objective being, obviously, the improvement of corporate performance.

The evolution of the CIO represents one of the more advanced expressions of the humanistic development of a company, and the focus will be on a persona that today is at the heart, or cornerstone, of the attainment of corporate development in that sense. It could happen that through the evolution of organisations, the systems of the future would assume different configurations. Until now, however, with scientific culture pervading principal organisations, the CIO seems to be the best qualified when it comes to sectors like Research and Development, or Human Resources and Organisation. More than others, the CIO possesses the right tools today. Still, he must absolutely change his own approach to his work and competencies.

The humanistic aspects of the CIO profession, and the activity of his/her company, are also related to the VUCA (Volatility, Uncertainty, Complexity, Ambiguity) perspective. VUCA is taken from the military linguistic code and applied for the benefit of the business. In the history of Western civilisation, mankind has experienced opposing philosophical approaches for interpreting situations and solving problems: the systematic/metaphysical approach versus the flexible/anti-metaphysical approach. The classic example of the first approach is Platonism and Aristotelianism, while relativistic and sceptic traditions are examples of the opposite approach. The advantage of adopting the flexible/anti-metaphysical approach is that one can avoid rigidity and fear of changing, which are obstacles in the evolution of any aspect of business and company, including the figure of the CIO. If the company embraces the flexible approach, which encompasses the awareness that any aspect of life—including business—is constantly changing and no form is universal and eternally stable, then the same company and its staff will be ready to face any unpredictable challenge and embrace serendipity.

Furthermore, in a certain sense we might say that the flexible approach includes the systematic one, while the opposite is not true. Indeed, the company adopting a flexible mode can decide to use a specific system for a certain period of time for the benefit of the company, with the awareness that the system can be changed or switched off any time, if needed. Switching from the "metaphysical" to "anti-metaphysical" approach is, above all, a mindset matter, a Weltanshanung ("vision of the world", to use a popular term from German philosophy), which involves redirecting any aspect of the company, including the role of the CIO. Once the company decides to embrace the VUCA approach, and the flexible knowledge and strategies it brings to the life of the company, the CIO might emerge as the key role to enhance and implement the new philosophy.

We believe that human sciences, or humanities, and a humanistic discussion practice within the company can help to increase the degree of awareness about the two available options "metaphysical" versus "anti-metaphysical", which is essential for deliberating the appropriateness of both.

We would like to stress the fact that the flexible mode does not necessarily imply dismissing any ethical approach to the exigencies of employees and clients. On the contrary, to balance the changeable strategy and the revolutionary/disruptive effects of it, the company needs to have a very strong and stable ethical agenda, which includes taking care of the mental and emotional wellness of the staff, a strong sense of responsibility towards the clients/final-external users of its service/product, a consistent communication, and a less-ceremony/more-participation mode within the company. Furthermore, the stress on ethical concerns should also be suggested for strategic reasons, since human resources play an instrumental role in the business. Moreover, these ethical concerns perfectly fit with the humanistic ideal we are proposing for the new CIO professional profile.

9 The Comprehensive Interaction of Cultures

As discussed in the previous paragraphs, we can distinguish between two types of organisational cultures: the scientific (or traditional) culture and the humanistic culture. The former adheres to a depiction of the organisational reality of a company composed of organisational charts, processes, tasks, responsibilities and roles, while the latter, on the other hand, tries to depict the organisation as a complex reality, consisting of interconnected flows of communication, collaboration tools, human resources that interact, the corporate climate, the psychology of relations, personal skills, and pervasive knowledge.

We represented the traditional culture as a "simplification approach" used by organisations to deal with the complexity of real-life corporate structures.

The traditional approach has been applied in corporate organisations for several decades and, despite recent technological developments, still prevails. Traditional IT tools were born in that environment. In information systems, this culture developed into corporate practice, introducing the concept of the "culture of the mainframe," with all the consequences that can be ascribed to it.

The humanistic culture was introduced later on, and has manifested itself in companies, including Italian ones, at different moments in time: we all remember the Olivetti experience, which introduced the humanistic notion of honouring the rights of workers to have the opportunity to dedicate time to their families even during working hours, and to have a work station conforming to standards of order, cleanliness and aesthetics. Other companies went ahead with projects like the Lean Organisation, aiming to reduce the number of hierarchical levels, delegate more and augment pro-activity, creativity, and innovation among personnel, thereby transforming staff into internal entrepreneurs. But rarely, or perhaps never, have these initiatives become an integral part of corporate culture, nor have they succeeded in effectively transforming it. One of the causes for this might be that we have always relied on the few initiatives at the higher hierarchical levels, without succeeding to implement a genuine change in culture.

These rare initiatives failed at the moment they had to be integrated into the cultural background of the company—an operation that must not be carried out in a simplified, and therefore scientific, mode, but with the tools of the complex management of change. Even if the best intentions were there, most projects did not create any value within an organisation, and some have even drained resources from training and testing—accomplishment-focused activities—without consequent practical results in terms of operational change.

The cause of the failures is connected to a specific methodological error: both cultures, the traditional and the new one, should not be considered at odds with each other, like in a manichaeistic approach, and they should not be implemented separately; they should rather be considered two faces of the same medal: they must be handled conceptually with the method of "comprehensive interaction," never by "substitution".

For the reasons described above, implementing a new system of collaboration must be carried out along with a redesign and simplification of the theoretical processes. The logic of comprehensive interaction must permeate these innovative projects and drive the change towards the new situation in-the-making.

As we already said, it might be necessary to employ rhetorical strategies in communicating with users so that people are collectively driven towards innovation without experiencing a conflict between traditional and new approaches. In other words, one must convince the users that the direction towards innovation is what they truly want. This technique was used not only by the most popular ancient sophists, such as Gorgias of Leontini and Protagoras of Abdera, but also by their major adversary, the Greek philosopher Socrates, who, with his pupil Plato, used a "noble" sophistic rhetoric to serve the good aims of the new Platonic philosophy against the traditional culture.

Also, the projects we define as humanistic must be carried out according to traditional modalities, though not exclusively so. The "Bimodal" approach proposed by Gartner is an expression of this concept and is increasingly becoming common practice. This means comprehensive interaction between project methods, the scientific approach (Waterfall) and the humanistic approach (Agile).

10 The Evolution of Competencies and of the Cultural Level of Staff: The Federal Organisation

The application of new technologies brings about, as noted, an upheaval in IT practices within companies, and the goal of these transformational projects is two-fold: (1) to redesign the cultural mindset at the Company, with the aim of increasing productivity and organisational efficiency; (2) to redesign the working modalities of people, reducing manual and repetitive tasks, leaving more time for analytical activities regarding operational processes, operational decisions, the empowerment of collaborative tasks, and the reinforcement of a collaborative environment.

This evolution of the scientific organisation into a humanistic organisation must take place, as we discussed above, through an integrated, rather than alternative process. The humanistic approach within an organisation should complement, rather than substitute, the traditional one. Competencies, too, must be integrated. For example, the competencies of synthesis of coordination, managing relations, and organisation of work must be integrated with the competencies of analysis, team work, empathy, conflict management, participatory leadership, motivation.

This cannot be accomplished exclusively by traditional training methods, or coaching activities. New training initiatives need to be developed and tailored to the personal strengths and aspirations of the people involved. Training programs for implementing this change must also be developed, especially for those who manage the resources in question (typically HR management), and perhaps they also need to be incrementally spread among the heads of the organisational architecture.

By freeing up resources and prioritizing new skills and competencies, the actual application of new tools will be possible in a comprehensive way, as will the execution of challenging projects with challenging targets.

This itinerary is extremely complicated and requires complex reasoning, analytical skills and knowledge of socio-behavioural dynamics. One could describe this trajectory of change by using a metaphor of a trail through obscure woods. We walk slowly, unaware of what is surrounding us, in pursuit of an outcome that only a few know to be there, and even they have no idea of its practical applicability. After our slow hike, we begin to see a dim light, and finally, as the darkness is thinning, our view of the final objectives gains shape and clarity. It is like reaching a clearing (i.e., our project and its goals) illuminated by the sun, that suddenly becomes clear to all who walk with us, and not merely the leaders.

But our trail doesn't finish here, as we must walk on another darkened trail towards another sunlit clearing, and so on. Indeed, the world outside continues to change, the technological opportunities revolve around us at the highest speed, the market offers new products and demands new solutions. And once more we take off in a cycle of continuous research. The search acquires a pattern, though, and a clear value becomes manifest: the culture and knowledge of the people involved. Knowledge and human collaboration as means to increase the quality of our community, even the business one, are humanistic values to pursue in the future.

Raising the level of knowledge means that an organisation can push on towards further delegating and autonomy. Maybe the functions that are more strictly operational must remain subject to major procedural rigidity, and to major directional leadership, but this is all part of the integrated model for the development of this process. We could add that, probably, this rigidity will be necessary only during the initial part of the journey, because subsequently all segments of the organisation will evolve and adapt to the new corporate mood. Through this process, the company will be thought of more and more as an organism instead of a machine.

The CIO will play the role of guide through this change: he/she will be a genuine Change Manager, or Innovation Manager. He/she will have the usual, traditional technological competencies as well as his/her new competencies, while the traditional distinction between Technology and Business, will be rejected in favour of a more unified vision. The future CIO will be much more integrated in the business and thusly be moreable to change it. He/she has to understand how to track an itinerary through dark woods to reach the clearing of knowledge, not only for him/herself, but also for all others. In our vision the CIO will be more a leader than an Officer; humanistic organisations will no longer need Officers, but rather guides towards the future evolutions.

11 Conclusions

This chapter has aimed to promote a humanistic type of evolution for organisations and for the corporate personae that would have to direct a transformation. We have attempted to understand what impact such a re-ordering may have within a company and whether it might somehow have an influence on productivity and profitability. Furthermore, we have addressed the necessity of a cultural change for achieving a new approach to working. Moreover, we have addressed what has been happening around us in terms of technological evolution and new opportunities offered by IT.

A constant concern in our discourse has been the relationship between the traditional scientific approach and the new humanistic approach to business and corporate organisation. We highlighted the complex dialectic between those two modalities and the necessity to adopt a "Bimodal" approach to avoid the negative effects of a disruptive evolution that occurs too quickly.

We also addressed the necessity of a new rhetoric and style of representation of the company system, no longer described as a machine but rather as an organism in which changeable situations, flexibility, emotional factors, and wellbeing of the "organs" must be taken into account. A total representation, indeed, in which people and the humanistic aspects of the business are at the centre.

The new role and features of the CIO, as we presented, reflect this big change. At the same time, we stressed that the company in which the CIO works has to understand and support his new role and the transformation he promotes. We also argued that the big changes the CIO promotes might scare the employees and the users inside and outside the company; therefore, it is particularly important to facilitate the company's internal communication to show the advantages that the new system can bring to all the subjects involved.

Because with great power comes great responsibility, the CIO has to fully understand the importance of his/her new role in the company. Accordingly, he/she will be more oriented towards leading the whole company with a new vision and mission, beyond providing the typical IT services. The core of the humanistic revolution through IT requires a CIO focused on human beings, as they are at the centre of the information network, devices, and practices he/she organises and leads.

This is an exciting moment and a great opportunity for the re-ordering of company organisation.

While we cannot fully predict what these changes will bring, we can decide now to take a chance, embrace the evolution with the means offered by IT, and accept the challenge of a new humanistic business.

References

Adorno TW, Horkheimer M (1973) The dialectic of enlightenment (German revised edition 1947) English translation: Allen Lane, London
Blosh M, Burton B (2017) Take a human-centered approach to digital design. Gartner research
Campus S (2014) La rosa nel calamaio. In: Van den Bossche Bart, Bonciarelli Sarah (eds) La collaborazione artistica nella letteratura italiana del Novecento. Franco Cesati Editore, Firenze, pp 131–137
Cain MW, Gotta M, Rozwell C, Create (2016) A business manifesto for digital workplace success. Gartner research
Decastri M (2016) Progettare le organizzazioni. Guerini Next
Fabbri TM (2010) L'organizzazione: concetti e metodi. Carocci
Ismail S, Malone M, van Geest Y (2014) Exponential organizations. A singularity University Book
Mason G (2010) Intranet 2.0. Tecniche Nuove
McMullen L, Papegaaij B, Meehan P (2016) Digital humanism requires an AGILE culture. Gartner research
Meehan P, Prentice B (2015) Digital humanism makes people better, not technology better. Gartner
Nunno T (2015) The wolf in CIO'S clothing. Gartner inc
Orban D (2015) Singularity. Hoepli
Papegaaij B, Meehan P, Prentice B, Olding E (2016) Leading from the heart: you don't become a digital humanist by fixing problems. Gartner research
Prentice B (2015) Apply digital humanism to customer experience design. Gartner
Prentice B (2015) How to apply gartner's digital humanism. Gartner
Raskino M, Waller G (2015) Digital to the core. Gartner inc
Spaltro E, De Vito Piscicelli P (2007) Psicologia per le organizzazioni. Carocci
Scholtz T, Prentice B (2016) Connect people-centric security to the digital humanist manifesto by starting and ending with people. Gartner
Varanini F (2015) Macchine per pensare. Guerini associati

The New Relations Among Things, Data and People: The Innovation Imperative

Dario Castello, Gloria Gazzano and Giovanni Vaia

Abstract In the first part of the book we argued that "people-centricity" must be a priority in the redesign of processes and operating models through digital technologies. A "people-centricity" approach, guided by the future CIO, is critical when a company digitally transforms people's daily experiences and approaches to business. In the following chapter, we present ways to leverage the human potential to structure internal relationships and manage external networks in order to drive the digital transformation. This chapter introduces the reader to the challenges posed by digital technologies as they design new relations among things, data and people. To fully exploit the digitally enabled opportunities, particularly process and business model innovation, we must consider the enabling factors such as capability design, digital innovation, environment design, internal organization design and digital IT governance.

1 Introduction

Today, digital technology enables the collection and analysis of data transmitted by multiple smart devices. Technology is creating entirely new ecosystems[1] with various stakeholders, including makers of tracking devices, security operation

---

[1] An ecosystem is defined by Skilton (2015) as "a connected convergence of technologies in a market and business activity that enable new consumer, business, and market performance and user experience".

D. Castello
Fiat Chrysler Automobiles, Turin, Italy
e-mail: dario.castello@fcagroup.com

G. Gazzano
Italgas, Milan, Italy
e-mail: Gloria.Gazzano@italgas.it

G. Vaia (✉)
Ca' Foscari University, Venice, Italy
e-mail: g.vaia@unive.it

© Springer International Publishing AG 2018
G. Bongiorno et al. (eds.), *CIOs and the Digital Transformation*,
DOI 10.1007/978-3-319-31026-8_7

centers, data analysts and other third parties providing value-added services. The ecosystems include new players as well as information-based items and information flows. Additionally, traditional players are also experiencing new roles and new challenges.

Let us look at a recent example from the insurance industry. In 2005, Unipol, one of Italy's largest insurance companies, worked with Octo Telematics, a large telematics provider in the insurance and automotive market. Together, they developed the first telematic policy in Europe (Unibox), installing devices in customer vehicles. Unibox included a 10% discount on premiums covering accident damage and a 50% discount on premiums covering theft (Vaia et al. 2012).

The system integrates OBU, GPS, and GSM technologies to capture and transmit location data, driving data, crash data and theft data. Many participants benefit from these information flows (trip data, policy data, cartographic data, crash data, theft data), as they can access data and reports online. The insurance company can collect data on millions of vehicle trips every 2 km.

Unipol and Octo have created a totally new ecosystem of services for different players: sales services to support new insurance policies; data entry of new contracts to initialize the service network; administrative services to charge and bill new policies; customer services for customer management and contract management (information and support on policies); road services to support end users during trips; behavior service to improve driving skills and car performance. Clearly, telematics has provided an important boost to service innovation and has had a significant impact on Unipol's business model (Vaia et al. 2012).

This ecosystem restructuration creates opportunities for new value innovation due to the realignment of data, function, and services. In Unipol's case, different stakeholders have joined forces to design the technology, share information and work as a dynamic meta-business system to build a valuable asset—without having to merge. But the full benefits of information-intensive technology investments have not been instantaneous. The timing of these enhancements is tied to extensive organizational learning and gradual consumer acceptance.

Thus, the availability of a vast amount of data pushes companies to **rethink how they create value for customers and how they capture that value**. Realizing new value depends on organizational learning and adaptation, involving many stakeholders, in a long **innovation journey within ecosystems**.

Incumbent firms in the automotive industry, for instance, are tackling the potential decrease of the market share due to the proliferation of new models, such as car sharing, and new competitors like Uber and Lyft. At the same time, big players outside traditional industry boundaries continue to show interest in the automotive business. Tech giants such as Google, Techstars and Amazon have set up research and innovation centers in Detroit to speed up this innovation process, without merging with car makers. Last year FCA and Alphabet signed an agreement to integrate self-driving capabilities with in-depth manufacturing capabilities, representing another key example of the collaboration between Detroit-Silicon Valley. General Motors has invested more than 500 million dollars on cruise automation technology by acquiring a San Francisco start up—Cruise Automation;

Ford invested 182 million dollars in Pivotal, a startup mobile application for cars and mobility. Others, like Mercedes-Benz, are creating new mobility services and business models: Mercedes Boots organizes transportation for children from home to school, sports or leisure activities. Therefore, the continuous knowledge exchange between engineers, designers, developers, and managers is creating a new industry.

In reshaping the boundaries of the business, digital leaders aim to govern innovation through a structured approach to the ICT Digital Transformation. This structured approach guarantees sustainable long term results, where people are at the core of innovation programs, **leveraging their creativity and innovation attitudes**, and **optimizing internal resources while facilitating external interactions**.

Naturally, the digital innovation journey needs to capitalize on the experiences of a multitude of actors, particularly those who are highly specialized in vertical solutions and who have the ability to rapidly adapt to changes, like small organizations/startups. **Orchestrating** data, ideas and technology becomes critical to managing the system of connected players.

Then, **contamination** is different from early engagement in that it is a true **co-design** and **co-development** and requires availability and willingness from all business partners that make up the ecosystem to play an active role during the overall digital transformation journey.

We present here three cases, ITALGAS, John Deere and Lago, that created and used new ecosystems to leverage company and market innovation potential.

Whilst ITALGAS improved their own capabilities on innovation just by integrating and balancing internal and external resources, John Deere and Lago respectively developed practices to orchestrate data, things and people in the system and co-design innovative solutions jointly with main stakeholders.

2 The Art of Balancing IN and OUT at ITALGAS

ITALGAS, a leading natural gas distribution operator in Italy and the third in Europe, is part of the Smart Energy & Utilities industry. Energy companies are constantly looking for ways to compete in this market, differentiating themselves from competitors, increasing the efficiency of their operations, and lowering costs for the consumer.

The world energy demand will increase by 40%, gas by 50%, while the trend will be reversed for coal and oil (International Energy Outlook 2016). Digital technologies will be critical for a more effective energy mix, in terms of low fuel consumption, optimization of resources, elimination of waste, less environmental impact, and the enabling of automatic and remote fine-tuning, for the purpose of transforming energy data into new services.

ITALGAS initiated a transformation journey (within the SNAM Group and in collaboration with the holding company) to benefit from IoT technologies and connections. The process designed at ITALGAS is twofold: one unit focuses on internal actors and the latter focuses on external actors. They call the first unit "IN-OUT" because ideas are originated inside the organization and then challenged by the external ecosystem. The second is called "OUT-IN" and engages with external sources for idea gathering, while evaluating the fit with the company business strategy.

The IN-OUT stream leverages the creativity of people and their knowledge of the business. It is an informal filter on innovative ideas that create value for the company. Conversely, ITALGAS's OUT-IN stream continuously seeks to map open innovation sources in an attempt to identify opportunities that can bring value to the organizations.

Both units originate ideas and proposals that could potentially bring value to ITALGAS. Here, innovation is the result of a combination of creativity (that they define as structured), execution (ability to transform opportunities into real-life use, bringing value to the organization) and appeal.

The process started back in 2012 (see Fig. 1). At the very beginning, the main direct factor pushing the digital transformation was "Innoseeking". Innoseekers scanned the main innovation sources represented by market analysts to spot promising ideas. All people inside the organization are potential innoseekers. They continually scout digital transformation opportunities and share their ideas and experiences with each other. In addition, they interact daily with the external actors of the innovation process, opinion leaders, vendors, consultants, in the overall open innovation ecosystem. Moreover, they interact daily with business stakeholders to

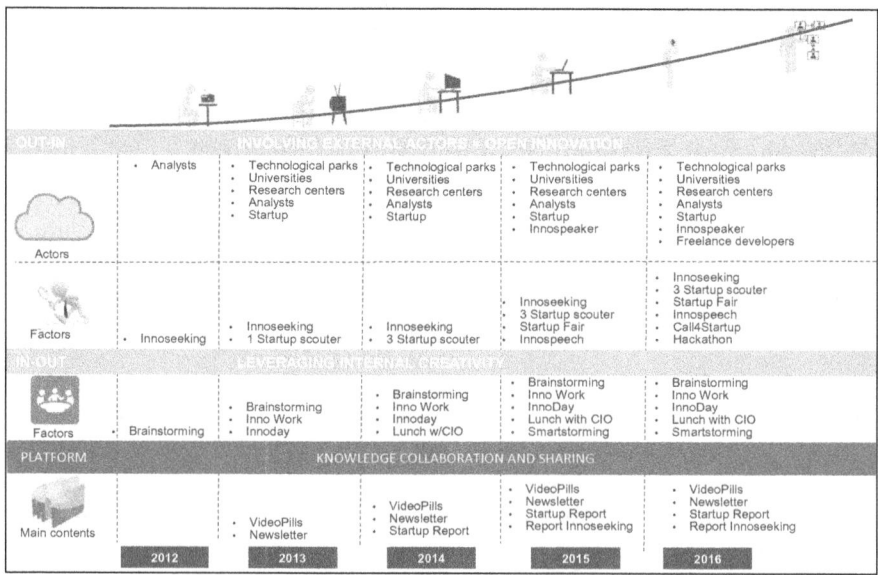

Fig. 1 ITALGAS's IN-OUT-IN digital transformation journey

collect continuous feedback on what is really critical and valuable for the business within a portfolio of ideas. Innoseekers are the real digital transformation engines at ITALGAS.

The process was formalized and supported by an innovation digital platform that was used as a document repository and document-sharing tool. The collaboration system consisted of internal actors, the innoseekers, and external actors (market analysts). A year later, they increased the number of external actors, adding Universities, Research centers, Technological Parks and startups (they introduced a formal startup scouting process). This became, over the years, the most relevant enabling factor from a process point of view.

Startup scouting is performed today on a recurring basis, and startups are assessed through a matrix model that evaluates the innovation level of the proposition and the applicability to the business. Startups enter in a funnel that usually leads to the development of a Proof Of Concept and eventually to deployment.

This scenario remained stable for a couple of years, then a new source emerged: the "Innospeaker". Innospeakers are visionary, subject matter experts, recognized by their community, for being engines of disruption. They play a key role in stimulating internal resources with elements of lateral thinking, helping them to think outside of consolidated schema. For this reason, ITALGAS encourages the involvement of people with very different backgrounds, even those unrelated to the business.

In parallel, ITALGAS further developed ways to engage startups. Contamination with the external actors is key to increasing the creativity potential of resources. To foster contamination, they introduced Startup fairs—events where all the company employees can meet the most promising startups and be exposed to their visions, prototypes, and solutions. Moreover, they use hackathons consisting of hands-on sessions and/or demos of solutions developed by the open innovation network, which can evolve into real products.

People's creativity coupled with their knowledge of the company business is the formula.

This formula is an important asset that needs to be developed as much as possible. To develop this asset, ITALGAS introduced three techniques:

Large Brainstorming: they decided to involve around 40% of the company's employees at each brainstorming round. The output of the brainstorming is a long list of themes to be submitted to the technical committee and to the Innoboard for evaluation. It is the responsibility of the Innoboard to convert this long list into a more actionable short list.

Smart Storming: Though similar to the Brainstorming, this technique is much more focused. It is guided by a facilitator that keeps the scope limited to a specific topic to be investigated and developed. It has been used as a sort of meta—methodology to define scope, priorities and tools associated with the innovation and digital transformation process.

Lunch with the CIO: These events are informal meetings where relevant topics from the scouting activities are shared with the CIO and other top managers of the IT organization. It can be considered an abridged version of the more structured opportunities selection process described before. But it is also a way to short circuit the strategy defined by IT managers and the resources needed to execute it.

Through this approach, ITALGAS accelerates the digital literacy of its employees and evaluates what brings value to the business. By having the technical and functional teams working side by side, they facilitate cross-fertilization and heavily reduce design time. They also build an open innovation ecosystem and expose their business to external knowledge and vision. As a mobility project to transform field operations, ITALGAS activated a network of almost 40 external actors (startups, research centers, universities) and peers, through a joint team of technical and business resources. Using "agile-like" techniques, the company supported the design and development of digital solutions: "continuous delivery of small functionalities is the best way to continually adapt to an ever-changing business scenario".

Finally, technology infrastructure plays a fundamental role in this case. ITALGAS's infrastructure is flexible enough to support fast, innovative and low cost solutions.

Today, they use a private cloud that has allowed them to cut management costs by 50% and has offered a provisioning time of about 15 min compared to the 10 days of the traditional approach. Management cost reduction eliminates entry/exit barriers and the automation of provisioning remove bottlenecks. This development is a great support for experimentation and innovation. The next step will be the adoption of a Software Defined Data Center (SDDC) to boost even more automation and support configuration processes. With a SDDC, they will be able to automatically reconfigure the Data Center infrastructure to provide computer resources to the applications just in time. So, when the business dynamics push for a digital service, ITALGAS will provide the required computing power; if the demand for service decreases, computing power will be rerouted to other services.

3 Orchestrating Resources at John Deere

John Deere was founded in 1837 by a blacksmith with a passion for inventions. Today, it is one of the key market players producing heavy machines for agriculture and green areas, with a presence in more than 30 countries. The mission of the company is to provide a set of reliable and safe agricultural and industrial tools to its customers across multiple business segments (industrial, agricultural, marine, retail and distribution, consumers). Nowadays, the company is the largest agricultural machinery producer in the world with a workforce counting 57.000 employees worldwide, and with a market value of $26.43 billion.

"John Deere has long been dedicated to those who are linked to the land, and is always ready to embrace change that leads to new opportunities," said Cory Reed, senior vice president of John Deere's Intelligent Solutions Group.

For the company, the need to increase production to face the world's growing population represents the chance to bridge the gap between a traditional sector, such as agriculture, with the era of connected devices, partly through the development of solutions on Smart Farming, namely Precision Farming. The Smart Farm uses a number of technologies, including GPS services, sensors, and big data to optimize crop yield. It is based on decision support systems that collect and process data in real-time, with the goal of providing information regarding all aspects related to the farming.

Today, Internet of Things is the enabler of precision farming, aiming at optimizing the efficiency and productivity of agricultural land, using modern and sustainable machines to get the best products in terms of quality, quantity and financial return.

Since 2011, John Deere has been developing a platform where customers can share data produced by their connected machineries, and then benchmark with other farmers and gain mutual benefits. The ExactMerge Intelligent Planter, for instance, is a tractor that places seeds at an exact space and depth, raising productivity and ensuring short planting. This 30 ton equipment system is guided by the Autotrac system, reaching an accuracy of roughly three inches. The same technology installed on sprayers permits a more efficient use of chemicals to protect from illnesses. Accordingly, sustainability of commercial farming can be increased and costs due to their excessive use can be reduced. The accuracy of the system allows reducing production costs in terms of labor, seed, fertilizer and fuel, with a significant improvement in operating efficiency and productivity per hectare. Furthermore, many other sensors augment the intelligence of these machines (the biggest possess 77 processors). JD Link remotely connects the machine with the office of the farm and mobile devices. Due to installed GPS sensors, the operator or farm manager can keep track of their fleet, monitor work progress, correctly manage logistics to avoid wasting time, and access important information like machine performance, speed, gas consumption, and early diagnostics. When smartphones, tablets and applications became more popular, the company upgraded its technology with the Mobile Farm Manager. More connected sensors were added to machines to enrich the monitoring capabilities accessible by those devices.

John Deere's vision for connected agriculture goes well beyond the individual farm. The final aim is to transform the agricultural industry by using data to push for collaboration between farmers and the entire ecosystem; a platform in which the grower can maintain close relationships with his trusted community of advisors, his equipment dealer, agronomic partners and also other farmers. Farmers and contractors rely on many partners and suppliers to carry out their activities (including manufacturers and distributors of fertilizers, software and agronomic services, consultants, etc.). Anyone can be a John Deere partner, just by adopting an "open system," and creating an interface for data communication from the Web portal to the backend of MyJohnDeere.com. The only requirement is to fulfill shared standards and integrity levels.

John Deere developed a new work management tool, which is integrated into the operations center and helps contractors as well as farmers to organize and do their job in an orderly manner and without using paper. Managers can access this function in two ways: from the office, as an integrated tool in the Action menu of the Operations Centre; or "on the go" using the appropriate app for tablets and smartphones. Machine operators will have their own version of the app "MyJobs," designed specifically to meet their needs.

A work management tool is valuable for contractors and farmers, as it integrates the whole process (work planning, organization of priorities, data logging, billing and reporting) and makes it much easier. They can share the type of operation, the customer name/field, information on the product use (for instance seeds/fertilizers), and the combination of required equipment. The presence of clear instructions eliminates the risk of misunderstandings. Furthermore, reliable and secure documentation substitute confused work reports drafted by operators. With a real-time overview of the machines and fields, the contractor/owner of the farm has a good level of flexibility that allows him to accept, plan and assign new orders. A calendar displays all the machines and the related tasks, to allow an effective planning several days, or a week, in advance.

Thanks to crowdsourced big data, coming from thousands of farms integrated with datasets on weather and other information, farmers can define the optimum levels of production. By connecting with retailers and buyers in real time, it is possible to optimize product delivery and transportation (Fig. 2).

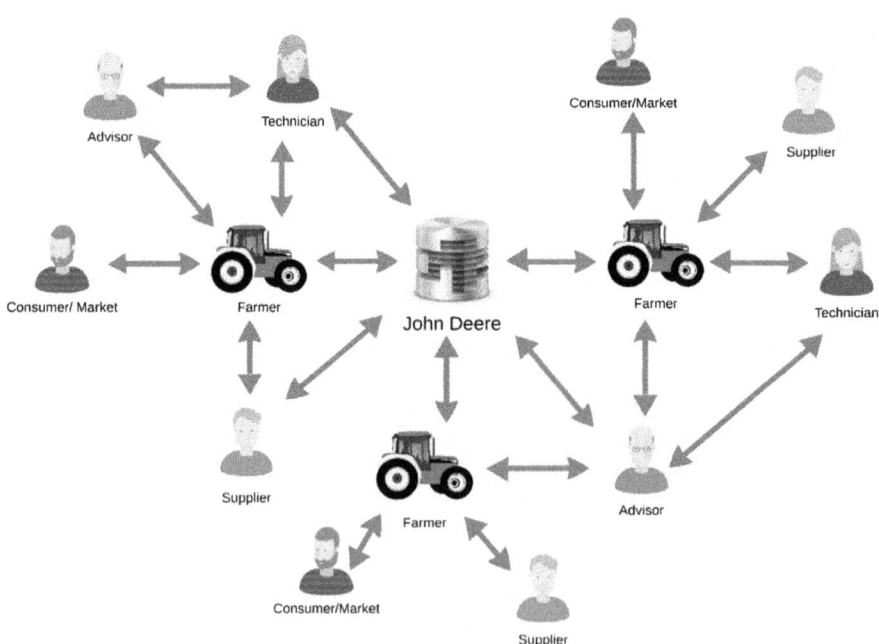

Fig. 2 John Deere agricultural ecosystem

John Deere's operating model has shifted from making 'big' machines to making 'smart' machines, through the creation of an information-sharing network made of equipment, sensors, satellites, and farmers. Over time, the characteristics of an ecosystem change. The keystone company should learn how to monitor the changes that the ecosystem undergoes over time. What gives the tractor an added value is the technology built on it, thanks to which the farmer becomes part of an agricultural ecosystem supporting his operative decisions.

The firm progressively redefined its boundaries, moving from a strategy focused on increasing production, to one aimed at the creation of a platform bringing together all the players of the sector.

In the future, John Deere will be able to act as an inspector of the performance of each company inside the ecosystem, an observer in critical changes and potential risks, offering reasonable suggestions and solutions to increase the effectiveness of a farmer's work. The business ecosystem becomes a surveillance network in which the IoT becomes the intelligent service platform that regulates the responses deriving from internal and external stimulation. The existence of such a rich community favors not only farmers but also John Deere itself, which will maintain close contact with customers and receive their feedback on a tractor's efficiency and usability, thusly leading to continuous improvements.

4 Lago as a Co-design Laboratory

LAGO S.p.A. is a company with a century-old tradition. It was set-up in the late 1800s by Policarpo Lago, who started his craft activities as a woodworker in noble villas and Venetian churches. His sons continued their father's activities and launched the production of design furniture.

During the eighties, the third generation expanded the firm and focused on furniture for living rooms and bedrooms. Since 2006, the company has continued growth in the high-quality design furniture sector under the guidance of Daniele Lago (today Chief Executive Office & Head of Design of LAGO). Under Daniele's leadership, the company revenue went from 5 million to 30 million euros in just 6 years (from 2004 to 2010), and it continues to grow year after year. LAGO is now present worldwide with more than 400 stores, both mono-brand and multi-brand, in big cities such as Rome, Milan, London, Paris, Madrid, Berlin and Prague. Moreover, LAGO furniture is present in many structures called "discovered," such as Bed and Breakfasts, resorts, hotels, offices, bars, restaurants etc.

The connection between products innovation, digital innovation and design is the secret to success at LAGO. "Interior life network" is the strategy that connects design and technology. It is related to LAGO's places (such as stores and public places) where the products can be bought or seen. Also, digital tools allow for the connections between the company and its customers and other stakeholders.

By employing the Interior life network strategy, LAGO has increased its turnover in the last five years from €24,23 million in 2012 to roughly €31,22 million in

2016. At the same time, the webpage sessions (index of users' interests in the company), the number of Facebook fans (index of users' engagement and brand awareness), and the customer contacts registered on the CRM (index of potential customers) have increased significantly.

Talking Furniture, an Internet of Things (IoT) project based on the NFC technology, is a great example of this approach that merges products, strategy and digital innovation. Indeed, this digital application is strictly connected with product and design development processes, and it is based on the growth of brand awareness, reputation and the deep involvement of customers. It is a highly innovative digital application that allows LAGO's furniture to be connected to people. The innovation commitment in the *Talking Furniture* development process can be considered very high; many development process tools have been used, such as a stage-gate process, to harmonize all activities across internal and external actors.

This effort in digital innovation has helped LAGO to also achieve high innovation performance, bringing it between "Stars" and "World Class" innovators.

LAGO establishes a special conversation with those interested in its design, and is enriched by the experiences and opinions of all participants. LAGO community counts more than 750.000 Facebook fans and 30.000 Instagram followers. The company is always seeking a mutually beneficial relationship with partners, such as architects/designers, the press, and customers.

LAGO knows its customers very well in terms of their preferences, passions, ages, economic situations, and geography. Social networks and Google are a paramount source of data for the company. By collecting and analyzing customer feedback, complaints, and recommendations, LAGO has been able to pinpoint a wide spectrum of customer desires, but also for fixing problems, developing complementary goods, and beginning to develop new products. This leads to a short development cycle period with lower development costs (Schilling 2010).

To reduce its development costs LAGO has adopted a parallel development process. With this method, some stages partially overlap, to encourage collaborations and interactions between internal departments and other actors. LAGO uses this method to minimize the length of the process and to align different people from different stages.

The management of new projects is in the hands of the digital marketing coordinator, who promotes collaborations and integrates all opinions (CEO included). He manages the projects from concept to launch, with the support of those responsible for each step (such as the graphic designer, web designer, developers, Information and Communication Technology managers, etc.).

During the innovation process, external people are principally required in two stages: scoping and brainstorming. Indeed, to generate more ideas and perspectives, LAGO organizes workshops dedicated to the creation of new product concepts. Product designers, selected from the best schools or architectural firms, generally take part in these workshops. Moreover, shop owners and Discovers' owners also participate, because they are always in contact with LAGO's consumers and can explain the design and development needs based on the customers' feedback. Workshops can take few hours, one day, or two or three days. All the ideas from the workshops are evaluated and screened, with only two or three of them moving into development.

To make use of all opportunities, efficiently exploit all available resources, and avoid wasting time, concept development and opportunity identification work together. It is very important that the collaboration involve decision-makers responsible for content, design, and development processes, due to the connection between customer usability and technical feasibility. For example, if a particular feature could dramatically improve the application but is too complicated or too costly to develop, two options are available: stop developing of that feature, or improve technical abilities to reduce the costs. Finally, the market launch starts with the product design stage, so that all communication activities are adequately prepared. Stage by stage, the strategy is refined. After the launch, the cycle restarts with new information provided by feedback and with new ideas to implement.

With *Talking Furniture*, the importance of relationships has been developed and extended even further. The company is developing its own social network, Memento, in which all LAGO customers can share experiences and moments of life around the furniture.

Memento is a function in *Talking Furniture* that allows people to record ideas, pictures, videos and texts in the NFC chip. In this way, all the activities and moments of life can be kept as memories in the furniture where they happened.

All these memories have three sharing levels. The first sharing system is the easiest one; contents are visible only to the person who recorded, so they are "private contents". The second level of sharing content is "at home," which means that the contents are visible to all those who activate the NFC on the product in which the contents have been uploaded. This level is useful for those who want to keep the pictures or texts only for a close circle of people. Finally, the third one allows sharing the contents with all LAGO customers. In this case, Memento gives two possibilities: sharing contents with people that have bought the same product or sharing contents with people that have bought any LAGO furniture. This is the starting point for the development of a real social network. People can review the products, can comment on pictures of other users and see in which city the furniture is.

5 Conclusions

Managing digital transformation requires a structured approach that involves an ecosystem of highly interconnected stakeholders. Our three powerful cases provide insight on how internal innovation is supported by a network of stakeholders (ITALGAS), how the IoT and collective data generation and exchange are used for better decision-making (John Deere), and finally how to crowdsource the new furniture concepts and the co-creation of content with social media users (customers) (LAGO).

These ecosystems perform well if external and internal contributions are organized and governed by a digital innovation leadership (other enabling factors, discussed in the next chapters, are needed such as a flexible infrastructure, the massive use of mobile devices, and a bimodal, that combines stability vs. rapid change, delivery mode). Hence, consumers evolve into "engines of innovation," increasing the innovation sources and innovation opportunities at a rate that is not manageable by a single organization.

These forces are changing the way organizations do business; they call the ICT function and its leader, the CIO, to action, to take command of this revolution. Indeed, complex relationships involve stakeholders in multiple business areas, and the "ecosystem" mode of work requires developing a new methodology. To change the methodology, one should be familiar with the "point of departure" and with the way ICT used to work prior to the transformation. Furthermore, some of the relationships within the ecosystem are interpersonal but others involve relationships between connected devices. To make this work, the knowledge of technology is indispensable and this knowledge typically resides within ICT.

We believe there are three main reasons for appointing the CIO as the leader of this Digital Transformation process:

1. The digital transformation is pervasive in nature and usually impacts multiple business processes and multiple organizational lines. ICT, by its own nature, is equally pervasive inside the organization and, in many cases, acts as "guardian" or documented source of business processes. ICT knows which processes can be impacted by the digital transformation (and how) and thus exploits opportunities; in this process, issues are evaluated and risks are eventually mitigated;
2. Most "Digital transformation initiatives" once mature, move to an implementation stage. Indeed, ICT must master the tools and methodologies for managing different types of projects for scale, costs, timing, and risks. Traditional methodologies need to be integrated with agile IT methods, usually held within the ICT function;
3. The third reason is of course the technology per se. We are dealing with a technology-enabled transformation: each new product and service, or business model, is supported by software or hardware. Any new model of a vehicle, for instance, embeds much more software than the previous model, and the additional software often enables the reconfiguration of the functionalities of the product. ICT is now a strategic partner inside any organization. Moreover, IT consumerization and the wide diffusion of digital technologies call for a critical integration with traditional legacy technologies in organizations. The CIO has an understanding about the "big picture," and how to make different technology components work together.

The CIO has the knowledge and the right connections to design an organization that encourages the participation of all our resources (business lines, other CXOs, technology vendors, ...). The CIO role is to facilitate and manage these interconnections.

At ITALGAS, the CIO and its first line of managers, within the INNOBOARD team, are responsible for defining the IT driven digital transformation strategy and represents the first gate for approval of IT digital initiatives. Here, the INNOMANAGER reports directly to the CIO and s/he is responsible for all operational activities related to the digital transformation program. S/he is usually a member of the Application Development department, because a good knowledge of business processes and services is a requirement. S/he is the orchestrator of the innovation processes inside ITALGAS.

The CIO can actually drive the Digital Transformation, through a structured approach to innovation management, introducing the right methodologies and tools, leveraging the open innovation ecosystem and putting people, both users and designers, at the core of the process.

We have in mind a multi-faceted role to drive and govern this transformation:

1. CIO as evangelist, to achieve a general awareness about digital transformation opportunities
2. CIO as digital planner, to design and lead a roadmap for the digital transformation
3. CIO as facilitator, to establish productive links between ICT, the business and other critical sources
4. CIO as integration lead, to facilitate integration of technology in business strategy.

S/he needs to search, support and develop to effectively exploit the potential of digital technologies, internally or externally. This nurturing should be practiced in a changed environment, where the organization has endorsed a flexible, semi-autonomous, horizontal ability to foster the agility and deep knowledge of small teams. In the end, a new environment should be created with new ways of thinking and new ways of dealing with customers, an environment where CIOs must orchestrate innovation across functions and external networks to reinvent and structure new value delivery models, combining and harmonizing physical and digital. Finally, digital initiatives must be well integrated into a unique digital company, where the implementation of a Digital Governance plays a critical role by supporting the change of behaviors and a decision-making culture.

The next chapters provide practices, cases, experiences and knowledge about the design and organization of a digital transformation journey.

References

Schilling M (2010) Strategic management of technological innovation. McGraw-Hill/Irwin, New York

Skilton M (2015) Building the digital enterprise: a guide to constructing monetization models using digital technologies. Palgrave Macmillan, Basingstoke

Vaia G, Carmel E, DeLone W, Trautsch H, Menichetti F (2012) Vehicle telematics at an Italian insurer: new auto insurance products and a new industry ecosystem. MIS Quart Exec 11: 113–125

Digital Capabilities

Daria Arkhipova and Carlo Bozzoli

Abstract Digital technologies are fundamentally changing the way companies operate—so much so that the notions of "digital transformation" and "digital technologies" have become nearly synonymous. Yet, there is another dimension to digital transformation—a human one—that seems to have been largely overlooked despite being of the utmost importance. Whether a company decides to shift to cloud-based software or engage in data-driven decision-making, the success of implementing the new technologies will ultimately depend on how rapidly the employees can learn to work well with them. Leading organizations are becoming increasingly aware of the fact that merely introducing new digital tools and instruments will not be sufficient for a successful digital transformation unless people inside the organization feel empowered to use them. The question then becomes: what are the core employee competences and skills that a company should nurture and develop to exploit the potential of digital technologies to the fullest? Can these competences be developed internally or should companies attract them from outside? How do companies identify employees who are willing to change and inspire others to engage with technology? In this chapter, we will shed some light on how digital technologies are reshaping work as we know it as well as look at how the Italian energy conglomerate Enel SpA is revolutionizing its approach to building the digital competences of its employees.

The views expressed in the paper are those of the author and do not necessarily reflect those of the company.

D. Arkhipova (✉)
Ca' Foscari University of Venice, Venice, Italy
e-mail: daria.arkhipova@unive.it

C. Bozzoli
Enel S.P.a., Rome, Italy
e-mail: carlo.bozzoli@enel.com

1 Digital Technology Trends

The tremendous explosion of digital technologies has created a demand for new competences and skills that were previously unheard of. Think about Chief Marketing Officers (CMOs). Gone are the days when marketers reached out to a consumer through TV, radio or press ad campaigns and then measured their effectiveness based on sales figures. The advent of new digital communication mediums, including mobile apps, web and social media, along with the availability of large amounts of highly detailed customer data have changed the very core of what a marketing professional does. Nowadays, successful marketing experts master an entirely different mix of skills, from creating engaging social media content to advanced analytics techniques.

Now think about Chief Information Officers (CIOs). Today, as more and more companies move their IT-related services to the cloud, the role of CIO within an organization gradually shifts from one of a service provider to one of an important strategy and technology partner. Today's successful CIOs are strong business-savvy IT leaders who rely on their solid technical knowledge to address business needs and drive digital innovation. In their new role, CIOs are at the forefront in understanding how emerging technologies can be applied to their companies' offerings and how to develop innovative customer solutions. This new role requires a highly diversified skill set that would allow CIOs to intermediate skillfully between business and technical teams within an organization and understand the language and dynamics of both worlds.

What are the technologies that change the way we work? Most organizations are already experiencing disruptions in their daily operations caused by mobile, big data and cloud computing. Still, when it comes to more recent digital technologies such as internet of things (IoT) and artificial intelligence (AI), many "traditional" companies outside the field of information technology still tend to think of them as part of some unrealistic science fiction scenario. Even though one would need a sort of crystal ball to predict with certainty the direction in which these technologies are going to evolve, the following five technological developments are believed to have become emblematic of the new digital world:

- Mobile
- Big data
- Cloud computing
- Internet of Things
- Artificial Intelligence.

Mobile technology is an umbrella term used for technologies that run on portable "mobile" devices such as smartphones, tablets and wearables. Mobile devices are sometimes referred to as "pocketsize computers" as they provide their users with functionality and connectivity at the levels comparable to those of traditional desktop PCs while at the same time being much smaller, lighter and more convenient to carry around. With the introduction of high-speed mobile internet, more and more consumers are starting to use their smartphones as a one stop source for web browsing, communication, entertainment, shopping and payment.

As our society becomes more mobile-centric, companies need to adapt to the changing patterns of consumer behavior and progressively shift towards mobile communication channels. In times when customers are used to having instantaneous access to any information anywhere and anytime, developing a mobile app becomes an imperative for companies that want to keep their customers engaged with their product or service. When it comes to workplace, mobile technologies have enabled employees to access information and communicate with co-workers outside regular working hours and off-premises. Today's professionals expect a great deal of flexibility when it comes to when and where they work: they may start their day by checking emails from their personal tablet while still at home, edit a file on their smart phone during their morning commute and then continue working on the same file on their office laptop. Efficiency-oriented companies are thus challenging the conventional assumptions about how office workflow should be arranged and are considering new approaches to work organization that would better exploit the potential of "employee-facing" mobile technologies.

Big data is a term used to define massive volumes of data coming from various sources within and outside an organization. Although traditional business intelligence input (e.g., transactional data on sales value and volume) counts as a part of big data, a much larger chunk of the data that actually makes it "big" comes from the variety of digital sources such as web, mobile and social media. The power of big data, however, does not reside in its volume but in the speed and quality of its decision-making processes. As companies get access to more data, the whole principle of business analytics gradually changes from a descriptive analysis based on historical figures to making informed decisions based on real-time predictive analytics. Put simply, big data analytics help companies to gain insight on why a certain pattern is observed and what can they do about it.

One example of a company that has improved its operational performance through intelligent use of big data in the energy sector is Enel. With more than 80% of its infrastructure digitalized, the company integrates historical performance data on its power stations with real-time sensor-based information on their operating conditions. Applying predictive analytics methods to the data has allowed Enel to timely identify potential issues and prevent failures based on what happened in the past to this same (or similar) infrastructure object under the same (or similar) operating conditions (Hirtenstein 2015).

Storing and processing the unprecedented amounts of data would have been impossible without *cloud computing* or, as it is frequently referred to, "the cloud". Cloud technology has allowed companies to access and manage terabytes of data over the Internet through third-party service providers, such as Amazon Web Services (AWS), without incurring large up-front capital investments into their own on-premise IT server infrastructure. Cloud-based infrastructure services (IaaS) are typically delivered on-demand on a pay-per-use basis, thus driving the operating expenditures down and making cloud computing resources accessible to large and small businesses alike. In addition to cloud-based data storage, businesses get access to cloud-based applications—or software as a service (SaaS)—that completely eliminate the need for installing, updating and maintaining software. Instead of installing software on a single physical machine, a user subscribes to a service online and enjoys full-time access to cloud-based applications remotely through web interface from any connected device. Furthermore, platform as a service (PaaS) allows customers to develop, test and deploy applications in the cloud without having to invest into software, web hosting and server infrastructure.

Though at first glance "the cloud" seems to be an entirely IT-related matter, it affects all employees across the organization as the technology allows them to access their work-related information from any device. The work becomes more transparent and co-workers can collaborate in real time from different geographical locations. Employees no longer need to worry about losing their data as backup happens automatically and files are easily recovered. Yet, even though accessing data and enterprise applications from personal devices undoubtedly increases employee engagement and helps them to get things done quicker and more efficiently, doing so may expose employers to numerous security threats. Even digital-savvy device users sometimes remain unaware of the risks that, say, irresponsible usage of third-party technologies may potentially entail and what needs to be done to prevent them.

As more organizations progressively move their software and infrastructure systems to the cloud, CIOs are expected to guide organizations through this transition and bring to the table their solid understanding of cloud computing technologies. They are required to restructure IT-related operations and gradually abandon legacy IT systems in favor of new technologies. As more business applications will be provided by third-party partners, assisting businesses in reviewing technical proposals and making technical due diligence will become invaluable. Moreover, even if migrating to the cloud reduces the workload that was previously related to operating server infrastructure and maintaining the software, the focus of IT shifts towards cybersecurity, disaster recovery, data storage and backup.

The advances in cloud computing have enabled the networks of connected physical devices to exchange data over the Internet—a phenomenon known as *the Internet of Things* (IoT). Such smart "things" are embedded with sensors and are uniquely identifiable through an individual IP address. This "connectivity" allows them to receive, register and transmit sensor-based information and, in some instances, perform an action remotely induced by an incoming signal. The idea behind it is not entirely new. Companies have used embedded sensors and wireless device communication (e.g., RFID, NFC) for more than a decade now, but the recent development of the underlying technologies such as cloud computing, mobile internet and miniaturization of sensors has brought what was known as machine-to-machine interaction (M2M) to a whole new level (Burris 2014).

The real value of the Internet of Things comes from the large amounts of data that these objects generate. This makes objects not only connected, but intelligent. Analogously to how a human brain learns from life experiences, artificially intelligent systems rely on *machine learning* techniques to automatically detect patterns in the data. The more observations are fed into the learning algorithm, the better the system gets at predicting a particular outcome. Think about Enel using machine learning for its predictive infrastructure maintenance. By processing a rich volume of historical sensor-based data on equipment failures and environmental conditions, it became possible to "train" the system on predicting power generation asset outages and to improve its accuracy over time. Similar principles of machine learning are used for fine-tuning speech and image recognition software, for training software bots to assist people in performing computerized tasks and for helping self-driving cars to navigate in controlled environments.

The aforementioned technology trends are not separate—they all build upon and reinforce one another. Working with these technologies requires new sets of IT skills and advanced technical knowledge in several domains. Take the example of big data jobs. While some of the skills required for big data roles—such as statistics, math and programming—are not necessarily new, these jobs do require a certain degree of familiarity with novel applications that allow storing, processing and manipulating large sets of data. If a job requires machine learning expertise on top of that, then experience with simulations, computational modelling, neural networks and learning models is highly desirable as well. Likewise, as mobile and software are "eating the world", the demand increases for mobile developers, software engineers and UI/UX experts (Gerber 2016b). As more mobile "things" get connected between themselves, the market need for experts with knowledge of device networking standards, electrical engineering and network security will continue to grow.

Does this mean that only highly specialized technical skills are in demand in a digital world? Not necessarily. As new technologies continue to emerge, today's "hot" tech skills will eventually become mainstream. If this happens, individuals will acquire the ability to learn new skills, problem-solve and logically reason what will help them to stay ahead. People with technical backgrounds may have an advantage in this regard as they develop a so-called "coding mindset"—an ability to break a problem down into small parts without losing the holistic picture of how these parts should work together as a whole.

But while the "coding mindset" quality is invaluable for tech roles, it may not be as important for other positions. Let us return to the Enel example. When assessing the digital competence levels of its employees, the company did not measure everyone according to the same standard. Instead, Enel combined tech and non-tech evaluation criteria, identified best performers for each criterion and developed personalized learning and development paths to reinforce employee strengths and help them to meet their professional aspirations. This made sense: a brilliant coder may not necessarily have an entrepreneurial spirit and out-of-the-box thinking, just as creative personalities might not always have the patience to spend hours trying to find a bug in a code. The key take-away from this example is that companies wishing to augment their in-house digital talent need to customize their learning programs to strengthen individual employee profiles and build a solid foundation for the continuous learning of new skills, once the current ones become obsolete.

2 Digital Workforce Outlook

Digital technologies have important implications for modern society and workforce as they redefine the way work is currently organized in most organizations. As companies continue on their digital transformation journey, they are becoming increasingly aware of the skills deficit they need to address in order to move forward (Manpower Group 2015). Consequently, organizations are experimenting with new ways of building digital competences in-house and are trying out alternative models for sourcing digital talent from outside. Digital technologies are driving societal and cultural changes, too. Many companies are gradually coming to the realization that traditional work arrangements based on full-time employment, office co-location and fixed working hours may no longer reflect the needs of their employees and their business in general. A new digital worker requires a more open, collaborative and dynamic work environment designed to unleash his or her full innovative and creative potential. Moreover, advances in cloud computing, high-speed mobile internet and connected IoT systems have jointly contributed to the development of sophisticated artificial intelligence algorithms that can outperform humans at certain tasks. With digital technologies progressing so fast, what will the workforce of the future be like? We have identified four tech-driven workforce trends that have already started to appearing many companies as they embrace digital technologies. We believe that the workforce of the future will be characterized as mobile, multi-skilled, on-demand and augmented.

2.1 Mobile and Distributed Workforce

As mobile connectivity and cloud technologies become commonplace, employees can access enterprise applications and data from anywhere, and the need for their physical presence in the office is reduced. Although remote work is not feasible in all organizations due to the nature of certain jobs (e.g., nursing, delivery), it has a greater potential for office jobs. For these jobs, employees can eventually be valued and rewarded based on the quality of output they have produced, regardless of the amount of time spent in the office.

Team work is facilitated since online collaboration tools and enterprise social platforms allow for interacting and soliciting advice from other team members. The boundaries of organizations expand because team composition is no longer limited to those present in a given location. Companies can access different skills and optimize their workforce at a low cost.

Information flows are becoming more transparent both within and between different functional teams. Previously, information flows were organized in a top-down fashion—to know what was happening in a different functional area of an organization, employees needed to wait for an update from their superiors. Now, the progress of other teams is traceable online and lateral communication between different functions is facilitated. Similarly, newly added members of a team are able to see the history of all prior project-related communications in a shared folder in an enterprise repository.

To sustain a distributed workforce model, workers need a certain skill set that allows them to operate and share ideas in a virtual world. The basic ICT skills, such as being comfortable at operating different hardware devices and interacting with business applications through user interfaces, are must-haves even for entry-level jobs. Since corporate content becomes accessible through personal devices, employees need to use their devices in a responsible manner and take active measures to protect one's device (e.g., regularly update anti-virus software, password-protect applications, automatically lock device when not in use, avoid insecure internet connections).

Employees must be familiar with the additional functionality of the remote collaboration tools that allow them to edit and synchronize files, set up video conferences, share screens and customize access settings. Digital technologies make work processes more transparent, make it easier to track the real-time progress and reduce the incentives to shirk. It does not mean, however, that team members are in "free flow". To manage workflow effectively, employees need to make use of online project management tools that allow them to set and communicate priorities, track latest versions of files and "flag" the tasks that have been completed. Doing so will make work more effective and ensure that members of a team are in tune with recent updates.

In addition to hard skills that make virtual collaboration run smoothly, employees need to be aware of behavioral consequences in a digital environment. In some cases, virtual co-workers never even meet each other face-to-face, and the online reputation that one creates becomes the sole basis for how one is perceived by his or her co-workers in a digital environment. Creating and managing one's own digital identity professionally (or maintaining several digital identities simultaneously) starts with controlling the type of personal information that is shared with the public. Employees are thus advised to customize their privacy settings, keep track of their digital footprint as well as respect other people's privacy.

Finally, following the established digital etiquette becomes essential for creating a positive work environment. Although much of how a person interacts online stems from the personal communication style he or she uses in "the real world", there are certain rules of business conduct online that one should adhere to. Some of these rules are common knowledge while others are more subtle. For example, when using an email for business communication, employees are advised to double-check the list of recipients before sending an email, write clear and concise subject lines, use bullet-points when sending a long body of text and make use of "out-of-office" notifications. Similarly, when using instant messaging, employees are invited to avoid long discussions, be mindful of their coworkers' availability statuses and respond in a timely fashion. Making employees aware of these simple rules of digital communication will make online work efficient and productive.

2.2 Multi-skilled Workforce

New technology developments have created the need for a variety of new skills and new roles, and many companies face difficulties in closing the talent gaps and filling new positions (Bessen 2014). An increasing reliance of businesses on real-time data in decision-making has naturally spurred the demand for "hard" skills pertinent to data mining and extraction, database management and analysis. The newly emerged job roles such as data scientist or data analyst have called for professionals with proven experience in and knowledge of large dataset analytics (e.g., RapidMiner), programming languages (e.g., Python, JavaScript, PHP), and computational and statistical software (e.g., R, SAS, SPSS, MatLab). Not only has data availability resulted in the creation of the new roles, but it has also generated the need for technical skills in the areas that have traditionally required soft skills. This is particularly evident for jobs in marketing and PR that used to be more about creativity and artistic expression. Even though soft skills such as "an eye for design," user experience, good writing and visual storytelling skills are still invaluable for digital marketing experts, creation of high-quality digital content requires knowledge of the functionality of technical tools for content production (e.g., Wordpress, Photoshop) and, in some instances, even basic coding skills (e.g., HTML5). At the same time, the marketing profession becomes increasingly about data analytics (Field et al. 2015). Today's marketers are required to identify patterns and trends in large datasets, quantify the return on investment by using web analytics tools (e.g., Google Analytics, Tableau) and experiment with and test creative digital ad campaigns (Gerber 2016a). Doing so requires understanding and applying the principles of statistics and math to be able to collect, transform and analyze data as well as interpret and draw meaningful conclusions from the results.

As data permeates almost every aspect of organizational decision-making, the increasing demand for technical skills comes as no surprise. What is noteworthy, however, is that soft skills become just as important as technical knowledge for IT professionals, especially in leadership roles. Strong expertise in IT is necessary but no longer a sufficient condition for CIOs to succeed: in addition, their roles require skills such as empathy, service orientation, negotiation, communication and team management. A company may develop great technology products but this alone will not help the business to succeed unless the IT executives understand the business environment, are able to put themselves into the shoes of their (possibly) non-tech users, empathize with their problems and clearly communicate the benefits of the solutions their company is proposing. Marrying "art and science", striking the right balance between creative and analytical thinking, having "both sides of the brain"—whatever the terminology, the mix between hard and soft skills lies at the core of most occupations in a digital world (Vozza 2016).

Another important challenge is that the composition of skills required for a particular job position is constantly changing and skills are not always transferable across different companies or industries. For example, according to the Future of Jobs Report, the skill sets required for the "data analyst" job role in financial services and consumer retail are very similar (World Economic Forum 2016). Conversely, there is very little overlap between what a data scientist is required to do in market research as opposed to energy industries. The mere rate of change and cross-industry differences imply that the reliance on formal job descriptions may be misleading in the digital age, and job titles in the future are likely to be defined as "agglomerations" of skills. Put simply, in times when skills get obsolete so quickly, people should be evaluated more based on what they know and potentially can do, and less based on what they have been doing in the past and what the job title for it was (Golden 2016). Since one's old skills and capabilities may not necessarily be the same that are needed for one's future role, the most valuable employees are those who have the right aptitude, and whose intellectual curiosity pushes them to develop new skills on their own. To solve the skill gap problem, companies thus need to focus on identifying and attracting "versatile" candidates that can adapt to the fast technology pace and are willing to learn continuously.

2.3 On-demand Workforce

Just as companies need to adapt quickly to ever-changing digital realities, so does their workforce. With the pace of technological change so frenetic, however, employees' skills are quickly becoming obsolete and even the fastest learners find it difficult to keep up with the new technologies. As a result, internal skills mismatch becomes an issue for many businesses. According to the survey conducted by Capgemini Consulting, 77% of companies consider the lack of in-house digital talent as a hurdle to successful digital transformation (Capgemini Consulting 2013). Besides, businesses often need to access specialized expertise only for a limited period of time and on an occasional basis. In these instances, a temporal need for a very specific skillset may not fully justify all the time and effort invested in a traditional process of candidate search, selection and recruitment. To find the right talent and gain fast access to rare competences, many enterprises turn to talent crowdsourcing platforms.

There is no exact operational definition for what crowdsourcing actually means. In fact, there is not even a single term to define this phenomenon. Call it crowdsourcing, or "gig" economy, or contingent, or "liquid" workforce—the core idea behind it is to access and leverage the potential of untapped talent pools outside the company "walls" (Accenture 2016). But if, in the past, crowdsourcing work tended to be associated with attracting low-skilled cheap workforce for executing routine "clickwork", today's freelance crowd on platforms like Upwork or Freelancer.com consists of professional web and mobile developers, web and graphic designers, writers, consultants, marketing experts—that is, highly qualified professionals craving cognitively rewarding and creative tasks (Soffer 2016). Thanks to digital technologies, freelancers were given access to software, tools and educational material to further develop their skills and do their work independently. The emergence of digital talent platforms made it possible for freelancers to showcase their work and reach out to employers all around the world.

Companies seem to become more aware of online talent platforms, too. According to Workforce (2020) report, 83% of executives around the world rely on non-payroll, contingent workforce in addition to their full-time employees (Workforce 2020 Report 2014). As new technologies are constantly emerging, contractor-staffed work arrangement is becoming a viable solution that allows companies to access specialized skills and deep expertise without incurring the costs of hiring or re-training a full-time employee. The "talent cloud" is especially relevant for startups and small enterprises that have limited financial resources, but it is equally important for the larger enterprises that start "scratching the surface" of new areas of technology and operation where they might not have sufficient in-house competences yet.

Lack of in-house competences is not the only reason companies use talent platforms. More often than not, companies turn to online communities to find new, fresh ideas. Take the example of General Electric. GE's aviation engineering team was struggling with designing a more lightweight metal jet engine bracket without compromising its mechanical properties (Stinson 2014). The team had a general understanding that the solution lies in using 3D additive manufacturing but they could not figure out the exact way to do it. In 2013, GE launched the public challenge of redesigning the engine bracket on GrabCAD, an online community with more than a million members with backgrounds in design and engineering. Several months later, after reviewing more than 1000 submitted proposals and testing the short-listed designs, GE announced the winner. The best solution offered 84% reduction in the weight of a bracket and came from a young Indonesian engineer with zero aviation experience, M Arie Kurniawan. This successful experience set an important precedent for GE's subsequent open innovation initiatives and collaborative projects.

A more recent example comes from the energy sector. In 2015, Enel Green Power (EGP), a subsidiary of Enel Group, launched a series of ideation "challenges" through InnoCentive, an online "marketplace for ideas" that allows corporate organizations to crowdsource solutions from private experts in a wide range of disciplines. With a focus on renewable energy sources, EGP was primarily interested in obtaining early-stage technology solutions and innovative ideas for preventing ice formation on wind turbine blades, assembling solar panels automatically or using drones during construction, and operation and maintenance activities in its power plants (Carmichael 2015). The company obtained more than a hundred different proposals and awarded a prize of €10,000 to each of the seven winners based on the technical feasibility, implementation potential and the idea's originality.[1] And, at the time of writing, Enel Group has three other R&D

---

[1] "The seven winners of the Innovation Competition of Enel Green Power", www.enelgreenpower.com, July 27, 2015. Retrieved from: https://www.enelgreenpower.com/en-GB/innovation/innovation/concorso_innovazione/.

challenges under evaluation on the InnoCentive network. By harnessing the "global brainpower" and complementing in-house R&D efforts with innovative thinking from outside, forward-looking companies such as Enel are able to resolve their challenges faster and accomplish better results.

2.4 Augmented Workforce

There has been a heated debate regarding the extent to which the rapid development of advanced digital technologies will lead to displacement of existing jobs and occupations. The fears of those who share a negative view regarding future employment issues are not completely unwarranted. Much like factory assembly line production eliminated a large part of the need for manual and physically demanding labor in manufacturing, now robotics and artificial intelligence systems seem to be posing a similar threat for knowledge workers and "white-collar" employees.

It is partially true. Indeed, most of the simple routine computerized tasks previously done by people (e.g., data entry, filling in forms, sorting email) are already being performed by machines and software bots. What's more, complex but mundane tasks such as data processing, information search and report generation will be increasingly handled by bots in the future. Several reasons explain this phenomenon. First, the sheer amount of available data that employees need to process makes it impossible for a human worker to perform the tasks at the same speed and with the same precision as the bots. Moreover, machines are fully rational when it comes to assessing risks and making decisions. Recent experiments have demonstrated that AI-empowered algorithms are as good as human experts in grading high-school essays and diagnosing eye diseases. But unlike human experts who can overestimate the likelihood of an event based on their most recent experiences or subconsciously favor a certain candidate, an algorithm will not let emotions and cognitive fallacies interfere with its decision-making process. Finally, machine learning—which is at the heart of artificial intelligence—allows software to improve over time and learn from its own mistakes. The human mind uses the same learning principles but it is always bound by an individual's past experiences and situations. Machine learning algorithms rely on large amounts of data and process millions of examples to identify hidden patterns and learn from them. Given this capability, it is not surprising that software has started to outperform humans at certain tasks. In fact, whenever a task can be described by a series of logical "if—then" rules, chances are high it will soon be replaced by an intelligent algorithm.

There is no denying that AI-empowered systems are now permeating knowledge-intensive professions—industries such as law, healthcare, finance and education—that were always thought to be immune to automation and impossible for machines to substitute. However, if we take a more nuanced view of automation in intellectual occupations, it becomes evident that different job tasks are susceptible to automation to a different degree. A recent McKinsey&Company study on automation technologies shows that very few occupations in fact will be subsumed by machines entirely (Chui et al. 2016). It is not the jobs but the tasks and activities that will be increasingly handled by technology. According to the study, jobs that are most susceptible to automation include large components of predictable physical work, data processing and data collection. That is, bots and machines are predicted to take away all the "boring" work—mundane and tedious tasks that are time-consuming, monotonous and prone to human error. This could include anything from information search and retrieval to appointment scheduling and administrative reporting. By "delegating" these tasks to a software bot, an employee frees up time for more fulfilling high-level work that helps to fully exploit his or her intellectual potential and creativity.

In addition to taking away all the "drudge work", bots can actually assist humans in performing their daily tasks faster and more effectively. According to Gartner research, an individual "pet AI" or, put differently, a virtual personal assistant (VPA) was listed among the top emerging workplace technologies (Pemberton Levy 2015). One of the most prominent examples comes from the legal industry. ROSS Intelligence has developed the first "artificially intelligent lawyer" that uses natural language processing (NLP) to understand and process the spoken questions and requests from its human colleagues (Alba 2015). The AI-powered lawyer saves days or even weeks that a human would waste to query legal databases and locate necessary documents. By taking out time-consuming and tedious components of the legal work, the software bot shifts the focus of a legal employee towards more value-adding tasks.

The applications of AI go beyond information search and retrieval. For example, by drawing on natural language generation (NLG) technologies, Narrative Science has recently introduced new software that is capable of generating a verbal description of a chart or a graph produced by the Tableau data visualization tool (Marr 2016). By generating explanations in a simple-to-understand language, the software facilitates the job of an analyst by gaining important insights from the data and communicating them to non-tech audiences. Artificial intelligence is also making its way into our e-mail services. Boomerang startup has recently launched "Respondable," an e-mail assistant that relies on the power of AI to help workers to write polite, actionable and informative emails (Finley 2016). If an email sounds too plain, or too rude, or uses too much negatively "charged" language, the software notifies a user and makes suggestions on how to improve it. All projects are still in their infancy but are perfect examples of human-machine collaboration in which AI-empowered bots are used either to "augment" user capabilities or to simply make users' life somewhat easier. At the same time, as technology continues to encroach on fairly cognitively demanding tasks, the nature of tasks performed by human knowledge workers will gradually move to a "higher ground" (Davenport and Kirby 2016).

Intelligent automation changes the demand and composition of skill sets required for performing a particular job. As machines start to handle many routine tasks and empower humans with insights and information, employees are expected to leverage their "augmented" capabilities and shift their focus towards the tasks that machines are incapable of doing. Take the example of customer interaction on social media. Since the early 2000s, brands were traditionally relying on human-to-human communication to engage with their customers on social media platforms. As intelligent bots and chatbots get more sophisticated, they will be the ones that respond to customers' technical questions and provide them with personalized recommendations. As this happens, human workers will need to readjust their skillsets to develop novel ways of engaging with customers and create more immersive user experiences using augmented and virtual reality (Edwards 2016). That is, much of the threat that automation presents is not necessarily about an individual being replaced by technology but about an individual who is not flexible enough to learn the skills that matter in a new digital environment.

A category of jobs that is predicted to be in high demand in the AI-powered environment relates to developing, supervising and maintaining automation software. The necessary skills will include identifying, selecting and optimizing work processes to automate. Moreover, one's ability to parse a complex work process into a series of logical steps and to define exception rules will become invaluable in the future. As intelligent as they are at solving problems that are readily presented to them, there is still a long way ahead until bots will develop the ability to recognize and formulate a problem, to understand the needs of another human being, to innovate and to discover alternative problem solutions—and these are exactly the skills that will be in high demand for the future jobs.

3 Digital Transformation at Enel

There has been a common belief that large incumbent players are facing intense challenges in spotting and exploiting the massive innovation opportunities that digital technologies offer them, and even more so in non-tech, slow-moving industries. The utilities industry is one of the best examples of that. For decades, companies in the utilities and energy industries have been operating as natural monopolies. Due to the nature of the industry itself, their efforts historically have been focused on ensuring service reliability and effective utilization of existing technical infrastructure, with little expectation of innovation.

But what used to be a stable and slow-paced industry is now undergoing fundamental transformation. On the supply side, companies are witnessing the growing importance of renewable energy sources and fast development of new energy storage technologies (Bocca 2016). As the Internet of Things technologies continue to advance, energy companies are being presented with immense opportunities to develop new products and enhance their decision-making thanks to data streaming through grid infrastructure. On the demand side, energy-efficient technologies and onsite energy generation possibilities are shifting energy consumption towards more sustainable and environmentally conscious modes. The entire competitive landscape is changing as well: tech giants such as Apple, Amazon and Google are starting to make inroads into the energy market with the intention to compete with existing energy companies in selling electricity to wholesale customers (Mulherkar 2016). These new players are aggressively innovating and have sufficient in-house competences to offer new technology-empowered services that the incumbents might find hard to compete with. In this uncertain and fast-paced world that utilities sector has now become, the existing players are pushed to foster the culture of innovation to be able to identify new opportunities and act upon them.

Like any other utility company, Enel was under pressure to transform its business in order to stay competitive in the digital era. A truly global business, Enel is a world-wide power manufacturer and distributor. With a net installed capacity of more than 89 GW and 1.9 million kilometers of grid network, Enel is able to supply electricity and gas to over 60 million customers in 30 countries around Europe, North America, Latin America, Africa and Asia, which makes it the largest energy company in Europe. By combining its unique scale of operations with an ability to pursue new opportunities in a connected world, Enel was determined to reshape the future of energy. The opportunities were there: any digital device needs energy to run, and Enel seemed perfectly positioned to connect the worlds of technology and power. The major challenge, however, was in overcoming the lack of digital "thinking"—the company did not yet have the right mentality to start disrupting traditional ways and reimagining the existing business model. There was a clear need for a culture that would stimulate innovative ideas and create possibilities for their rapid execution.

In 2015, a digital transformation strategy was launched by Francesco Starace, CEO of the Enel Group. The transformational program has three core "grand" objectives. First, to instill the culture of openness and innovation in people. Second, to increase efficiency in operating and managing company assets, generation and distribution networks alike. Finally, to develop innovative services and build a sustainable competitive advantage in new and mature markets. The Group's Head of Global Information and Communication Technology, Carlo Bozzoli, has been leading the part of the transformation aimed at tackling six major challenges of the global ICT:

1. **Optimizing the application portfolio.** As a result of Enel's long history of growth through international acquisitions, the company has a portfolio of over 1800 enterprise applications and almost one hundred different technologies. Given these figures, simplification appears to be an urgent matter—a smaller number of elements would be easier to operate, maintain and keep under control. Today, it is not only important to develop new applications but it is equally critical to simplify the existing application portfolio and reduce the number of technologies Enel relies upon.
2. **Transitioning to the hybrid cloud.** Leveraging cloud technology and adopting cloud-based infrastructure has represented a more robust, more flexible and cost-effective solution for Enel. A complete transition to the cloud, including but not limited to IaaS, is a very complex process whose implications go far beyond moving to a cloud-based platform. It entails an entirely new mindset that changes the way people within the company start perceiving IT, the way they

start developing collaborative DevOps-like approaches to work organization, and the way they start understanding the importance of organizing new processes in a nimble fashion. It leads to a culture that celebrates "continuous development and deployment" that, in turn, speeds up time-to-market and allows for bringing in customer insights at earlier stages of product or service development. The new work paradigm offers massive opportunities for the company's employees to grow professionally and reach their full potential, but it also requires new skills and extensive training for Enel's leadership team to make it work.
3. **Engaging with suppliers**. As a general trend, companies will be shifting from buying products to buying services. In response, Enel has been developing a new sourcing model that would better reflect the evolving style of client-supplier relationship and redefine the way suppliers are selected, evaluated and retained. As this trend continues, suppliers are expected to support the company during the entire product life cycle. The role of suppliers is changing from a passive task executor or technology provider to one of an important technology partner committed to results and highly motivated to making a positive impact during and after Enel's digital transformation.
4. **Developing ICT operating and service models**. To effectively promote and manage innovation, Enel has established a 'focal point'—a unit specifically dedicated to observing and identifying relevant technological trends. The unit needs to assess the degree of consistency of the selected technologies with the Group's strategy, their applicability and feasibility of implementation. On the internal side, Enel is actively working to digitalize end-user services and create a better user experience for its employees in their daily activities. For example, a unique IT service portal has been created alongside a series of digitally-enabled initiatives such as Global Service Catalogue, Self Help—Self Service, multi-device access, and multi-contact communication channels (chat, web-based tools, etc.). The company seeks to facilitate the adoption of innovative services by engaging and empowering Enel people.
5. **Fostering digitalization and innovation**. As digital technologies are permeating every aspect of people's lives, Enel is using digital tools on a massive scale to engage with its customers and make the most of in-house data. Enel has three priority concepts to make sure that digital services are truly adding value: "Think and act digital", "Communicate digital" and "Be digital". "Thinking and acting digital" means being able to use digital technology to improve the way employees work on a daily basis and reinvent legacy business processes. The idea behind "Communicate digital" is to put in place new tools for smooth and efficient communication—new intranet, web-based platforms and social media —to keep up with the most recent office communication technologies. Finally, the motto of the "Be digital" concept is "Technology is ready—but are people ready, too?" and it seeks to assess the ability of people within the company to become digital ambassadors within their areas of influence.

6. **Evaluating digital competences**. Enel has quickly come to realize that it makes little sense to invest in digital technologies if people are not ready to use them. But different people have different degrees of "readiness" depending on their personal experience with technology, their willingness to try out and learn new things, and their attitudes towards change and digital technology in general. To evaluate the extent to which people at Enel were prepared for a digital transformation and whether they could manage traditional and digital business models at the same time, Enel developed a competence assessment program. The program was named 6Digital and included multiple evaluation stages. The assessment starts with identifying "evangelists" throughout the world within the Enel Group. These are employees demonstrating strong digital skills as well as the desire to share knowledge and a creative view of the future. "Evangelists" are early adopters of technology that are particularly enthusiastic and knowledgeable about new digital tools. As their opinions and recommendations are generally respected among their peers, there is a greater chance that "evangelists'" positive feedback may encourage "technology laggards" to get out of their comfort zone and embrace change. The process identifies multiple types of evangelists, which are then asked to participate in Hackathons and reverse mentoring.

When talking about the transformation program, Enel's Head of ICT Carlo Bozzoli compares his team's taskto that of a "GPS navigator" that helps the business to find the way through the maze of potential opportunities and directs it towards the right ones. In Bozzoli's own words: *"We are trying to understand which processes and skills must be built internally and retained within the company, and which ones can be sourced from outside, using new mechanisms that were not available to us in the past such as crowdsourcing platforms, partnering with start-ups, universities, research centers, etc. We are currently also undertaking a process of "transformation factory" which provides, compared to the past, for strategic platforms, the insourcing of key competencies to better oversee the introduction, adoption and development of technologies"*.

The new mission of Enel's ICT is to enable the company to develop new business models and seize the opportunities offered by digital technologies. Nowadays, IT remains instrumental in creating the new culture of innovation and its role within Enel has been changing dramatically. Prior to the transformation, IT at Enel was a service provider, detached from the business-end and having little influence on the company's strategy and business development. Now, IT people at Enel work hand in hand with business colleagues at all levels. IT is actively involved in strategy discussions on equal footing with business executives. IT is a crucial member of any team working on the development of new business opportunities. In fact, when asked about the future of IT within the company, Carlo Bozzoli said he believes that the boundaries between IT and business will be getting more blurred until IT "dissolves itself" into business entirely. It represents a fundamental cultural shift within the company and the solid basis for the subsequent innovation. Indeed, one of Enel's first commitments was to entrench a belief in colleagues that every service of the company relies on solutions and technologies of which employees themselves are the primary users.

To further strengthen their agile and innovative mentality, Enel has been investing a lot of effort into developing the culture of open innovation within the company. Because the company has increased its reliance on external sources for ideas and talent, it developed a strategy for Open Innovation and devised new governance mechanisms to put it into action. The strategy is overseen by the Innovation and Sustainability function—a unit that has been specifically created at a Group level to manage activities related to open innovation globally. Innovation Committee was set up to track the progress of open innovation initiatives within the company; its monthly meetings are chaired by the Group's CEO and moderated by the company's Chief Innovation Officer (CINO). One of their most recent initiatives was setting up an Innovation Hub in Tel Aviv, Israel—a startup accelerator that provides industrial support and business mentoring to young entrepreneurs working on projects that "marry" technology and energy. In addition, an Innovation "In & Out" program was developed with the purpose to build and sustain collaboration with companies, universities, research laboratories, startup incubators and to "funnel out" the most promising partnerships.

Another new unit that was created to accommodate the company's needs during the transformation is called Digital Business Enabler. The unit's mission is to promote the development of digital business solutions within the Group, as well as to manage the relationships with partners and suppliers, when related to digital projects. The Digital Business Enabler unit prioritizes the initiatives, evaluates new opportunities of collaboration with other relevant units at Group level, and supports project managers assigned to the digital projects. The unit's activity falls into three major domains: Digital Services, Digital Communication and Digital People. The goal of the first one, Digital Services, is to ensure the smooth implementation and reliable operation of technological platforms on which the company's services are running. The second one, Digital Communication, aims at leveraging cutting-edge technologies and mobile solutions available on the market to enhance employees' internal and external communications. Finally, activities related to Digital People are focused on enhancing the digital skills of Enel employees.

Developing in-house digital competences continues to be of paramount importance for Enel. To remain competitive, Enel needed to become proficient in gaining insights from the large amounts of data that its infrastructure objects and customers are generating. Many business initiatives that Enel has recently launched are, in fact, relying on big data analytics. With the support of professional consultants and research labs, Enel has implemented predictive maintenance models for its generation power plants, renewable energy assets and its distribution network. Furthermore, Enel has partnered with innovative tech startups to develop analytical models for understanding behaviors and attitudes of "socially responsible" consumers. And even though collaborating with others has been extremely fruitful, Enel has always been aware that the partnering strategy alone is not sustainable in the long run if the company's internal digital capabilities are not being built at the same time. To successfully compete in the world of digital, Enel needed to combine external sourcing with building its own capabilities in-house.

4 Digital Competence Development at Enel

"Digital culture should be promoted and sustained from inside"—this was the central idea behind the 6Digital project launched at Enel Italy in summer 2015. This experimental project, aimed at fostering the digital culture within the company, initially involved Communication, Market, ICT and Innovation functions and, after Italy, was rolled out sequentially to Spain, Eastern Europe and Latin America. As the lack of "buy-in" from employees often stands in the way of any transformation, the primary purpose of the project was to identify employees with above-average digital skills so that later they could "evangelize" digital culture among their less technologically advanced peers. The project had also several positive "collateral effects" besides the officially stated purpose. First, conducting company-wide digital skill assessment provided an overview of the current level of digital competences within Enel and helped in identifying areas for improvement. Second, running such a large-scale initiative meant communicating to all employees that the company was committed to its digital course and was taking the change very seriously. Third, the project increased the awareness of the topic of "digital" within Enel and spurred the interest in those that had been previously doubtful about new technologies. Finally, the company was sending a clear message to potential hires outside. When it comes to choosing an employer, many highly-trained technical professionals prefer fast-moving technology-services companies to large enterprises in non-tech industries as they believe the latter cannot offer much in terms of professional growth and development. By investing in the digital competences of its people, Enel was signaling to the talent crowd beyond its "walls" that it was doing new, interesting things with technology and was offering promising career opportunities, *on par* with "young" technology-software companies.

Company leadership played an important role in making the 6Digital initiative happen. Even though the project leveraged the power of the crowds in promoting "digital thinking" within the company, it would have been impossible to accomplish without the top-down involvement of the company's leadership. Prior to launching the project, Enel's management first needed to have a clear understanding of what competences the company needed to move forward with the transformation. The type of competences required depended a lot on which digital technologies the company considered most relevant for its business and which of them had the most innovative potential in the mid- and long-term. In Enel's case there was, as the company's Head of ICT Carlo Bozzoli put it, "a triplet of digital technologies"—big data, cloud and mobile—that mattered the most as they have made the development and implementation of the Internet of Things possible (Teruzzi 2016). Hence, getting access to the technical skills such as computer programming, data analytics, cloud computing and cybersecurity have become a priority for Enel.

There are several ways to access those skills. A company might start scouting for new skills outside and partner with companies and people who have already mastered them. Too much reliance on external knowledge, however, puts the company at risk of not developing and not accumulating in-house knowledge fast enough to be able to spot emerging opportunities for innovation. Thus, partnering strategy must be complemented with internal efforts aimed at building one's own capabilities. Finding and recruiting those specialized skills is a viable solution that, however, can be time-consuming and expensive. On the other hand, knowing and understanding the competence level of the existing employees can increase the chances of finding a suitable candidate internally, even though his or her current job might not have required those skills in the first place. 6Digital program at Enel was developed to identify digital talent within the company.

The program envisioned a company-wide skill assessment that involved the entire staff of Enel. Putting the program in action was no small task since more than 70,000 people—employees of all levels, white collars and operational staff alike—were to take part in the survey. 6Digital project was first tested at Enel Italy and then was gradually rolled out to other locations. In essence, the project included three stages:

- *Digital Champions Assessment.* The first stage of the project aimed at identifying so-called "digital champions" across the organization through a survey. The survey, based on a proprietary model developed by a third party, consisted of two parts: digital readiness and a lateral thinking assessment. Based on the results of the assessment, all participants were divided into clusters based on their individual levels of technical expertise and creative potential.
- *Hackday.* Employees that scored highest on the digital readiness assessment were invited to a two-day coding marathon—Hackday. During the event, the participants were simulating a start-up: they were divided into teams and were expected to develop a mock-up version of a web- or mobile application based on the ideas that were proposed to them.
- *Digital Engagement Program.* After digital champions had been identified and their skills tested during Hackday, they were invited to participate in an 8-week engagement program. During the program, the participants deepened their knowledge in more advanced areas, learned how to effectively share their skills with other colleagues and contribute to building a strong digital community within the company.

We will now review each of the stages in more detail.

4.1 Digital Champions Assessment

The project started with an individual Digital Readiness Assessment (DRA) which was made via an online survey tool. The core idea behind DRA was that at the time of the assessment any employee may already have a good starting level of digital skills and the right attitude. Indeed, in some cases employees might have developed specific expertise as technology amateurs: for example, by engaging in activities such as blogging, programming, or managing online communities in their personal time. In other cases, employees may have a solid professional background in technology but simply have not been given an opportunity to apply their knowledge in their current role. The assessment thus allowed identifying those talented individuals.

Four criteria were used to evaluate to which extent an employee was "ready" to work closely with technology:

- Personal technology equipment
- Frequency of use
- Willingness to share knowledge and information
- Aptitude towards entrepreneurship.

Based on these criteria, participants were grouped into six categories:

- Analog native—these individuals are skeptical about the use of technology and are reluctant to any change in general. They exhibit very critical attitudes towards new tools, are unwilling to learn and prefer "good old ways that work". They may be difficult to influence and are unlikely to use new digital tools until they are forced to.
- Networker—people who fall into the "networker" category use technology in their professional lives but are followers by nature. They recognize the advantages of technology but are rarely passionate about it. They are often unaware of the potential of technology and use the basic functionality of digital tools that are offered to them.
- Digital star—"stars" are smart users of technology who believe that new tools simplify their life a lot. They use technology extensively both in their personal and professional lives for communication, information search and entertainment. They are well aware of both the opportunities and risks that come with new technologies and moderately experiment with the new tools to discover new features.
- Digital guru—for "gurus", technology is hobby and passion. They enjoy working with technology and have fun with it. Digital gurus are the ones that might learn a new programming language or open their own YouTube channel just out curiosity. They proactively search for new tools available on the market but are able to critically assess their quality and distinguish between good and mediocre products.
- Startupper—those classified as "startuppers" take technology very seriously and perceive it is a job. They are passionate about technology but are equally curious

about the business aspect of it. It is not enough for them to passively consume what the market offers—they feel the need to become experts in certain questions and propose their own new solutions.
- Hacker—what distinguishes a "hacker" from all other categories is his or her advanced coding skills. They are likely to master several programming languages and have experience with a suite of tools needed for developing user applications. These skills are obtained either through professional education or extensive self-training and are often verified by virtual communities of practice.

After the profiles of the "digitally ready" employees had been determined via DRA, Lateral Thinking Assessment (LTA) was conducted for those participants who demonstrated positive attitudes towards technology use. As the name suggests, LTA was aimed at evaluating an employee's problem-solving skills and his or her ability to think outside the box. Contrary to DRA, which put a lot of emphasis on technical expertise and hard skills, LTA was more focused on soft skills applied in a digital world. For example, the participants were evaluated based on their ability to construct non-obvious mental associations and propose creative solutions, their proclivity to experiment and take risks, their ability to write an original story and to use visual metaphors. Based on the survey results, participants were classified into four types depending on which personal qualities they manifested the most:

- Pragmatist—these individuals put a lot of emphasis on the practical application of ideas. They judge the quality of an idea based on whether it is applicable in real life and reject purely theoretical ideas as useless. Pragmatists rely a lot on their past practical experience in their decision-making; they like to take risks and act fast. Rather than spending a lot of time wondering about "what would have been", they test the idea in practice, observe the results and immediately act upon them.
- Specialist—those classified as specialists have a mind wired towards logic and analytics. They see value in experimentation, but make decisions only after all possible alternatives have been thoroughly analyzed. Specialists are great problem solvers: they invest a lot of effort in understanding the nature of the problem, gathering all relevant information and analyzing each possible solution in great depth.
- Methodologist—if "specialists" are great problem solvers, methodologists are great problem identifiers. They are very skilled at conceptualizing the phenomenon they have observed in the real life, spotting the patterns and developing abstract models and theories based on them. Their ability to represent any process as a series of steps makes them excellent planners.
- Creative—these are people with vivid imaginations that are able to think outside the box. They spend considerable amounts of time contemplating reality and making observations that go unnoticed by most people. Their ability to challenge the conventional ways of thinking increases their chances of coming up with a radically innovative idea. Patience and meticulousness might not be their strongest points but they make up for it with their creative energy, original thinking and artistic expression.

Employees with the most representative profiles were identified as Digital Champions. Once both assessment tests were completed and employee profiles were obtained, the project entered its next stage.

4.2 Hackday

Those employees that were identified as Hackers and Digital Gurus received an invitation to take part in a two-day coding marathon—the Hackday. The main goal of this event was to recreate the startup environment and boost interest in experimentation and collaborative approaches to developing new products and services. Hackday was organized as a team competition: the participants were divided in teams of six to eight people and were expected to develop a working prototype of a web platform or an application for one of the announced topics. For example, the teams of 6 digital Hackday in Milan in 2015 were assigned to one of the three categories. The first one, "Enhancing the daily life of Enel's employees", invited new digital solutions to simplify the way Enel employees perform their tasks at work. The second one, "Transforming energy use", called for creative ideas using digital technologies to change the way energy is consumed. Finally, the third one, "Consumer utilities daily life", encouraged the development of applications for consumer everyday use.[2] By the end of the second day, the teams needed to submit a working prototype, even if it was not 100% complete. Only mock-up projects with a functional front-end were considered and evaluated, whereas projects containing a simple idea description with PowerPoint slides were not admitted. The evaluation criteria included the project's compliance with the assigned topic; the quality of user interface and its visual design; technical feasibility of the project; technical functionality and completeness of a prototype. The winners for each category were selected by a jury.

At first glance, Hackday is a part of Enel's overall open innovation agenda. Indeed, the initiative was largely focused on sourcing ideas and solutions to company challenges from within and unleashing the creative potential of the employee "crowd". Undoubtedly, Hackday participants generated many fresh ideas that might have inspired the company's leadership to develop them further. However, the benefits of the Hackday go far beyond that. If digital readiness assessment has elicited Hackers and Gurus with above-average technical skills, the Hackday was the perfect environment to test if people can apply these skills to creating actual workable digital tools. The participants needed to demonstrate their knowledge of one or more programming languages and their familiarity with

---

[2]"Alla Scoperta del 6 Digital hackday di Enel". Enelsharing, November 11, 2015. Retrieved from: http://enelsharing.enel.com/innovazione-area/scoperta-6digital-hackaday-enel/.

development tools. Furthermore, Hackday was an important step toward building a digital community: gathering people with a similar "coding mindset" who share a common interest for programming in the same place increased their willingness to exchange ideas and continuously learn from each other.

4.3 Digital Engagement Program

The final phase of the 6digital program—Digital Engagement—served a long-term goal of creating a global digital community of people who were willing to advocate for new technology use within Enel. To that end, Hackers, Gurus and Digital Champions were invited to participate in an 8-week Digital Engagement program. The program included intensive training to familiarize participants with new technology-driven trends (e.g., ecommerce, Internet of Things) as well as enhance their soft skills for effective management, communication and mentoring in virtual environments (e.g., reverse mentoring, Lean Start up). The program was also aimed at encouraging collaborative spirit between the participants and increase their willingness to share skills and knowledge with their less technologically advanced peers. The "graduates" of the Digital Engagement program were expected to evangelize and reinforce the digital culture through informal communication and networking.

5 Lessons Learned

Enel is good example to how a company in a non-tech industry approaches the challenge of upskilling its workforce in a digital age. It should be particularly inspirational for other large companies in traditional, non-tech industries that may still be under the false impression that the digital revolution does not concern them. Large, established companies have the most important asset for succeeding in leading transformational change: their people. What these companies need to do though is to retrain them and prepare them for the challenges of the future, and here are some guidelines based how Enel has managed to do it:

- **Define strategic priorities first**. A company's leadership needs to have a clear idea of which technologies will create value for the business. The type of skills and knowledge required will depend on what things a company wants to do with technology.
- **Identify, test and upskill the digital champions**. Employees differ in their skills, knowledge and attitudes towards technology. Skills assessments help to identify high-potential candidates who have enough base knowledge and are willing to learn more about digital. Internal "start-up" competitions work well for testing their skills in practice and crowdsourcing ideas for company challenges.

- **Nurture the culture of sharing and collaboration**. Digital champions should not be perceived as an "elite club" within the company. Instead, they should share their skills with those that are less adept at using new tools and lead people by their own example.
- **Kindle the interest in those that are change-resistant**. People are more likely to try out new tools if they see value in it for themselves. Illuminating less digitally-aware employees about the benefits of technology and "taking them by the hand" in using new tools increases the chance of people actually adopting them.
- **Partner with experts**. Developing internal talent is important but not all the best people are already working for the company. Working with smart people from outside always brings in a fresh perspective and provides opportunities for knowledge exchange.

Such talent transformation endeavors are not easy and, in part, much of their success depends on execution. Indeed, HR and other dedicated units play an important role in pushing such initiatives forward. But most importantly, the success depends on whether the company gets a general direction of change efforts right, and this is when the CIO becomes key. One of the opening lines of this chapter was that today's CIOs need to complement their tech expertise with business knowledge. But if one zooms into what makes a true *digital leader*, it becomes clear that it takes a much broader mix of skills, knowledge and attitudes to succeed. Great digital leaders ask the right questions. They stay abreast of the latest emerging technologies and are able to distinguish between temporary hype and the game-changing trends. Digital leaders are highly knowledgeable about the specifics of their business, their customers and their industry and hence are able to confirm if a particular digital initiative applies to their company or not. They are audacious enough to inject innovative tools and to disrupt the established routines but they are risk-conscious and do not expose their company's employees, data and customers to unnecessary security threats. Finally, true digital leaders understand the importance of people in making digital transformation happen. They experiment with new approaches to work organization, give autonomy to their employees and invest a lot of effort in addressing competency gaps. And these investments in human capital generally pay off manyfold because—to quote the motto of Enel's digital transformation—"there is no digital strategy without digital people".

References

Accenture (2016) Liquid workforce: building the workforce for today's digital demands. Retrieved from: https://www.accenture.com/us-en/insight-liquid-workforce-planning

Alba D (2015) Your lawyer may soon ask this AI-powered app for legal help. Retrieved from: https://www.wired.com/2015/08/voice-powered-app-lawyers-can-ask-legal-help/

Bessen J (2014) Employers aren't just whining: the 'skills gap' is real. Harvard Business Review. Retrieved from: https://hbr.org/2014/08/employers-arent-just-whining-the-skills-gap-is-real

Bocca R (2016) 3 trends transforming the energy sector. World Economic Forum, March 2, 2016. Retrieved from: https://www.weforum.org/agenda/2016/03/3-trends-transforming-the-energy-sector/

Burris D (2014) The internet of things is far bigger than anyone realizes. Wired, November 2014. Retrieved from: http://www.wired.com/insights/2014/11/the-internet-of-things-bigger/

Capgemini Consulting (2013) The digital talent gap developing skills for today's digital organizations. Retrieved from: https://www.capgemini.com/resource-file-access/resource/pdf/the_digital_talent_gap27-09_0.pdf

Carmichael S (2015) Five challenges to drive innovations in renewable energy. InnoCentive Blog, September 8, 2015. Retrieved from: http://blog.innocentive.com/2015/09/08/five-challenges-to-drive-innovations-in-renewable-energy

Chui M, Manyika J, Miremadi M (2016) Where machines could replace humans—and where they can't (yet). McKinsey Quarterly, July 2016. Retrieved from: http://www.mckinsey.com/business-functions/business-technology/our-insights/Where-machines-could-replace-humans-and-where-they-cant-yet?xid=nl_daily

Davenport TH, Kirby J (2016) Just how smart are smart machines? MIT Sloan Management Review 57(3):21. Retrieved from: http://sloanreview.mit.edu/article/just-how-smart-are-smart-machines/

Edwards C (2016) Why and how chatbots will dominate social media. TechCrunch, July 20, 2016. Retrieved from: https://techcrunch.com/2016/07/20/why-and-how-chatbots-will-dominate-social-media/

Field D, Visser J, De Bellefonds N (2015) The talent revolution in digital marketing. BCG perspectives, September 25, 2015. Retrieved from: https://www.bcgperspectives.com/content/articles/marketing-technology-organization-talent-revolution-in-digital-marketing/

Finley K (2016) AI is here to help you write emails people will actually read. Wired, August 23, 2016. Retrieved from: https://www.wired.com/2016/08/boomerang-using-ai-help-send-better-email/#slide-1

Gerber S (2016a) 15 essential skills all digital marketing hires must have. Mashable.com, March 8, 2016a. Retrieved from: http://mashable.com/2016/03/08/15-skills-digital-marketers/#NSXXx7sH.uqR

Gerber S (2016b) Web developers, data scientists, AI experts: the 15 top tech roles of 2016. Mashable.com, January 20, 2016b. Retrieved from: http://mashable.com/2016/01/20/technical-hires/#r2ggklCa5kqE

Golden G (2016) Why LinkedIn should kill the résumé and replace it with the experience graph. TechCrunch, August 14, 2016. Retrieved from: https://techcrunch.com/2016/08/14/why-linkedin-should-kill-the-resume-and-replace-it-with-the-experience-graph/

Hirtenstein A (2015) Big data squeezing more power from wind signals next investments. Bloomberg, October 12, 2015. Retrieved from: http://www.bloomberg.com/news/articles/2015-10-12/big-data-squeezing-more-power-from-wind-signals-next-investments

Manpower Group (2015) Talent shortage survey. Retrieved from: http://www.manpowergroup.com/wps/wcm/connect/db23c560-08b6-485f-9bf6-f5f38a43c76a/2015_Talent_Shortage_Survey_US-lo_res.pdf?MOD=AJPERES

Marr B (2016) How automated narratives are now helping tableau users gain more insights from data visualizations. Forbes, September 2, 2016. Retrieved from: http://www.forbes.com/sites/bernardmarr/2016/09/02/how-automated-narratives-are-now-helping-tableau-users-gain-more-insights-from-data-visualizations/2/#6a7f09da7d2f

Mulherkar A (2016) Google, Amazon and Apple are forging the future of corporate energy management. Greentechmedia, July 8, 2016. Retrieved from: http://www.greentechmedia.com/articles/read/What-Google-Amazon-and-Apples-Recent-Moves-Reveal-About-the-Future-of-Cor

Pemberton Levy H (2015) Top 12 emerging digital workplace technologies. Gartner, October 8, 2015. Retrieved from: http://www.gartner.com/smarterwithgartner/top-12-emerging-digital-workplace-technologies/

Soffer P (2016) Understanding the economy of the crowd. TechCrunch, September 4, 2016. Retrieved from: https://techcrunch.com/2016/09/04/understand-the-economy-of-the-crowd/

Stinson L (2014) How GE plans to act like a startup and crowdsource breakthrough ideas. Wired, November 4, 2014. Retrieved from: https://www.wired.com/2014/04/how-ge-plans-to-act-like-a-startup-and-crowdsource-great-ideas/

Teruzzi E (2016) IT life: Per Bozzoli (Enel) l'IT deve diluirsi nel business. TechWeekEurope, April 28, 2016. Retrieved from: http://www.techweekeurope.it/projects/bozzoli-enel-cio-94168

Vozza S (2016) Why 2016 is the year of the hybrid job. FastCompany, March 15, 2016. Retrieved from: http://www.fastcompany.com/3057619/the-future-of-work/why-2016-is-the-year-of-the-hybrid-job

Workforce (2020) Building a strategic workforce for the future. Research results presentation by Ed Cone, Oxford Economics at SAP Success Factors, September 18, 2014. Retrieved from: https://youtu.be/a35BtvzS-CA

World Economic Forum (2016) The future of jobs: employment, skills and workforce strategy for the fourth industrial revolution, January, 2016. Retrieved from: http://www3.weforum.org/docs/WEF_Future_of_Jobs.pdf

Designing the New Digital Innovation Environment

Massimo Messina

Abstract If we think that the digital economy is uniquely driven by technology, we risk making a big mistake; to build a real digital business we need to integrate different ingredients. This effort could require, for example, putting the customers at the center of the business with complete control over what is happening around them, integrated logistics for the products that you sell, an obsession with quality, and of course, compelling technology. Still, technology is a true element of a new digital company and, in some cases, it is where the competitiveness and the originality of the products are created. Certainly, technology cannot be effective if it is not fully integrated into all the company's processes and its organization. It also requires adequate, flexible, and low cost Information Systems. The understanding of which type of changes should be applied to these factors is therefore a key element of any digital transformation plan in order to fuel a profitable digital business. The organizational matter also crosses another very important aspect: the scouting, farming, and transformation of the competences of the personnel involved in the new digital business. These competences include new technical skills as well as an understanding of what is needed to transform the current information systems so that they can be used to map the journey from where you are today to where you want to be in the future. In this chapter we are going to explore such questions as:

1. What are the trends in new technologies and how are they influencing exponential organizations?
2. How are new technologies, like containerization, different data organization, and processing algorithms setting the ground for a new way to manage IT in enterprises?
3. Is it possible to build systems with very high quality, zero downtime and a very strong resiliency against negative events?

The views expressed in the paper are those of the author and do not necessarily reflect those of the company.

M. Messina (✉)
Unicredit, Milan, Italy
e-mail: Massimo.Messina2@unicredit.eu

4. How should the approach to software development, testing and management be transformed?
5. What architecture should we use to build modular, elastic, adaptive, and much more efficient IT solutions?
6. How central is DevOps in the new IT world? Are Cloud and Utility Computing the same thing?
7. How are successful companies developing digital IT organizations that can both foster business innovation and assure robust and effective deployment?
8. Does one IT in our company still suffice or do we need at least three: transformational, traditional and autonomous?
9. How should the IT technical team be re-organized and reskilled to support this new development and support approach?

The chapter will introduce the reasons behind these changes and suggestions on what can be done to deal with this complex and articulated transformational challenge.

1 A "New Normal" Context for IT Solutions

In the "new normal" context, (Hinssen 2010) companies dithering on the opportunity to reinvent themselves as a platform for new ecosystems will risk finding themselves unaligned with current trends.

When a company decides to become a platform, it also decides to change its relationship with what is considered a "product"; the platform in fact extends the concept of generated value to include the direct contribution of external customers, who in turn create value to be consumed by others. This relationship is particularly present in ecosystems where the actors are simultaneously consumers and producers (prosumers), amplifying the effects of what we saw during the Web 2.0 phase and giving a new meaning to the word "openness". It must be understood that the fear of being open represents a paradox in the digital world and that "security" cannot be a synonym for limitations and closure. The technology layer must therefore enable this exchange of value through Open APIs that can be used to empower the functionalities of the platform and to permit easy interactions among all the involved actors. It is very important to note that, following this reasoning, the business layer must become open as well; this layer must coordinate the external behaviors and contributions so that they follow the company's business strategy.

The blurring of these boundaries also effectively ends the conceptual distinction between business and IT. The past concept of IT as a lever for business process automation does not match the new meaning of IT as a weapon to effectively compete in the new environment.

Moreover a comprehensive digital customer experience requires a full integration of IT capabilities in the business (products, client approaches, devices, …) and a cultural, managerial and organizational change at the enterprise level.

Designing the New Digital Innovation Environment

This paradigm shift puts new pressure on the application requirements, which have changed dramatically in the recent years:

- moving from domains where the companies had full control of functions and workloads to domains where the control is also outside
- moving from tens of servers, hours of offline maintenance, tolerance for unavailability, seconds of response time, and data measured on a gigabytes scale to continuous deployment, computing power on cloud, infinite scalability, zero tolerance for incidents and unavailability, any device, maximum performance in milliseconds, data measured on a petabytes scale
- moving from a rigid and indoor design with a time-to-deliver of months to a flexible even-driven, interconnected, API based design that with agility can achieve a time-to-deliver of weeks
- moving from silos to open architectures where interfaces are well documented (by APIs), open standards and open sources are used when it makes sense, and future layout changes are not a nightmare, but rather normal events to be delivered quickly and cheaply.

These requirements are influencing not only IT architectures (of both applications and infrastructures) but also organizations, processes and tools used to develop and maintain IT solutions; the next paragraphs will guide you through the exploration of most of them.

2 The Impact of New Infrastructures and Technologies

In the paragraphs below, we have tried to avoid giving shallow information about the important elements for innovating the context in which we work to make it digital; we have instead focused on their mutual influences and how these may change the perception of Information Technology usage. They will be given brief definitions only where the inflated use of the term might otherwise lead to multiple interpretations. To coherently address many items, we have grouped them in the following four categories:

- **Computing**—Cloud, Server, Storage, network, data centers
- **Applications**—how applications will change in architecture and development tools
- **Information**—Big Data, Machine Learning, Deep Learning
- **High-powered Peripherals**—IoE, Augmented Reality, 3D printing, wearables.

Each category also influences the other categories, thusly creating disruptive combinations that are feeding the exponential business.

2.1 Computing

In the technical sense, Infrastructure is the hardware, networking, and software that runs the foundation of an information system. This infrastructure includes fiber optic cables, hard drives, servers, switches, routers, firewalls and usually an operating system (Qrimp 2008). From a software developer's perspective, this infrastructure runs the code that produces screens, data, and workflows that the end users interact with; even in this perception, the infrastructure was still seen mainly as hardware. In a while, we are going to see how this notion is changing and how the distinction between hardware and software is continuously blurring.

By itself, infrastructure isn't useful—it just sits there waiting for someone to make it productive in solving a particular problem. Imagine an interstate transportation system. Even with all these roads built, they wouldn't be useful without cars and trucks to transport people and goods. In this analogy, the roads are the infrastructure, and the cars and trucks are the platform that sits on top of the infrastructure transporting people and goods. These goods and people would be considered software in the technical realm (Qrimp 2008). As for the interstates, it is much more convenient for us to pay a fee when we need to use a segment of the road, instead of paying a fixed cost for all the interstate network; likewise, it is really inefficient to have a private interstate, used only by one individual instead of having the multi-tenant infrastructure shared among all customers. Therefore, we are able to buy access to the infrastructure and pay a fee commensurate with our usage, without worrying about the theoretical limit of the infrastructure's capacity; we are transforming an infrastructure into a service.

The Infrastructure as a Service (IaaS) is a term used to acquire the capability to execute workloads on an infrastructure, without worrying about the physical footprint of the infrastructure itself (e.g. the number of servers being used at one time and what those servers are doing); the term also includes being able to dynamically expand or contract the amount of the resources involved.

Referring back to the metaphor of trucks and cars, recall that we compared them to platforms carrying the payload of the infrastructure. Well, instead of buying the cars and trucks, we could rent them, obtaining access to a service. Similarly, a Platform as a Service (PaaS) is one additional layer on top of IaaS that makes it easier and faster to immediately deploy applications without theoretically worrying about installing and maintaining the required hardware and software stacks. Obviously, we can still have a private infrastructure, but exploiting a public multi-tenant cloud could have better economical and quality figures, especially for the small and medium size installations.

> Computation is moving into the cloud, and thus into Warehouse-Scale Computers (WSCs). Software and hardware architects must be aware of the end-to-end systems to design good solutions. We are no longer designing individual "pizza boxes," or single-server applications, and we can no longer ignore the physical and economic mechanisms at play in a warehouse full of computers. At one level, WSCs are simple—just a few thousand cheap servers connected through a LAN. In reality, building a cost-efficient massive-scale

computing platform that has the necessary reliability and programmability requirements for the next generation of cloud-computing workloads is as difficult and stimulating a challenge as any other in computer systems today. [...] (Barroso et al. 2013).

In other words, building an IasS or PaaS environment for an Information System in a way that enables ANY application of the enterprise to run in a similar environment is no simple task, making standardization key. Indeed, the application architectures may require different types of hardware and software that may not be readily available as IaaS or PaaS, and thusly need excessive customization on the cloud services for installation, maintenance and operations; this problem could ultimately jeopardize cloud business cases. Therefore, this lack of a common standardized platform for the enterprise applications is a big inhibitor for some companies to adopt the cloud approach.

As professionals in some industries are already aware, technology is now mature enough to exploit the cloud capabilities even for complex ITs and more traditional and regulated business (Financial sector, for example), lessening the effects of the standardization problem: "Utility Computing" is not a dream anymore.

We can now supply computing power and storage like we do for electric power, paying for what we consume, without great worries (e.g. workloads, peaks, performance management, availability, disaster recovery, equipment obsolescence); such a system can dramatically change the costs and performance of IT departments, and positively impact the competitiveness of the company.

Just to be clear, the new Utility Computing is not a synonym of Cloud. For years, the cloud providers have increased the autonomy of their customers in the provision of rented environments. Still, for large enterprises the lack of standard architectures was making the customization of the cloud expensive and time consuming, thusly lowering the intrinsic benefit of being on the cloud. In fact, while for relatively simple Information Systems it was possible to draw ex-ante a set of standard platforms to be included in the catalog of cloud providers, in complex environments, the multitude of legacy applications with dependencies on old technology stacks (and maybe not operationally automated) still required a substantial amount of manual work that transformed the cloud contract into an outsourcing contract.

A new convergence of technologies is creating a discontinuity that fully enables the Utility Computing paradigm, making it highly exploitable even for large and complex installations. The baking ingredients for a modern utility computing are:

- Hyper-convergence
- Shared nothing pattern
- Reactive Systems
- Software defined infrastructure
- Container technology.

Integrated together, these technologies with their five nines availability, are able to sustain any legacy configurations (or make a migration business case positive) with any workloads in the new form of Utility Computing, except those using the old mainframe (but making modernization an option). Below, we'll explore each of them.

Hyper-convergence is a type of infrastructure, with a software-centric architecture, that integrates seamlessly with computing resources, storage, networking and virtualization and other technologies in a single box supplied from a single source.

A hyper-converged system allows integrated technologies to be managed as a single system through a set of common tools. Hyper-converging systems can be expanded to the base unit through the addition of nodes. The most common cases include virtualized workloads; in fact, we could think of this configuration as a data center in a box, in which you have everything needed to run an application.

Hyper-convergence was born from the concept of converged infrastructure. According to the Converged Infrastructure approach, a supplier provides a pre-configured set of hardware and software in a single box with the objective of minimizing compatibility problems and simplifying management. If necessary, however, the technologies in a converged infrastructure can be separated and used independently. Such a practice is not preferred, since the technologies of hyper-convergent infrastructures are so integrated that they hardly can be divided into separate components.

Due to the redundancy of these solutions and their pre-defined configurations that minimize change problems, they are very well suited for high availability services; moreover, their set of provisioning tools are a good introduction to DevOps.

The great advantage is that, typically, you can run legacy applications on a converged infrastructure with minimal changes, if any.

Much more invasive, from an applications point of view, is the concept of Shared Nothing. The term is a remnant of older terminology used in database architectures and is opposed to the other common infrastructure architecture: The Shared-Everything. The Shared Everything architecture refers to a system of architecture where all resources, like storage, memory, and the processor, are shared (Krishnan 2013).

Two versions of Shared-Everything architecture are Symmetric Multi-Processing (SMP) and Distributed Shared Memory (DSM). Today, this type of architecture has commonly been used in many transaction processing systems, where the transactional data is small in size and the resources consumption is consumed in short cycles and in the form of multiple processors sharing storage; for example a common database cluster is a "shared storage" system, since you can run multiple servers which all communicate their state back and forth over a network storage device (let us say NAS or SAN). This is easier to build, but the storage device may bottleneck, with the consequence that you cannot easily distribute the storage and computations on the cloud, implying also complex configurations to manage.

Shared nothing (SN) is a distributed computing architecture where multiple systems (called nodes) are networked to form a scalable system (Krishnan 2013). This means that neither disk nor memory is shared and all communication is done over the wide network (Stonebraker 1986).

Each node has its own memory, disk and processors and is independent from any other node in the configuration, thus eliminating contention and isolating each node from the others. The operating system of the node is managing its resources and not the application server. The flexibility, resiliency and scalability are the major strengths. The shared nothing architecture enables the creation of a self-contained architecture where the infrastructure and the data coexist in dedicated layers (Krishnan 2013).

The shared nothing also moves from single servers for each component (customized clustering solutions or high-end hardware machine/appliance) to a model in which several nodes contribute to providing the same service by splitting the workload amongst themselves. Any coordination between nodes is done at the software level, using a conventional network that handles failures and maintenance activities, such as adding or removing nodes, without downtime. Availability is guaranteed by a large redundancy and it is possible to reach a five nines availability level. The ability to split the workload in many independent servers (as is done in Hadoop) is one of the fundamental elements of elastic computing. In fact, by monitoring the need for resources, you can activate/deactivate nodes based on a workload demand; this characteristic is very important since, with Utility Computing, you pay only for what you consume.

The shared nothing approach thus opens the road to Reactive Systems. The original definition of what a Reactive System is can be found in the "Reactive Manifesto" (Reactive 2014), signed, at time of writing, by 16,000 specialists from around the world.

Many organizations in different industries are independently discovering patterns for building software that looks more robust, more resilient, more flexible and better positioned to meet the digital requirements that have changed dramatically in recent years. In the end, the need is for systems that are Responsive, Resilient, Elastic and Message Driven. The "Reactive Manifesto" calls them Reactive Systems.

Responsiveness is the cornerstone of usability; responsive systems focus on providing rapid and consistent response times: This consistent behavior, in turn, simplifies error handling (problems may be detected quickly and dealt with effectively), builds end user confidence, and encourages further interactions (Reactive 2014).

Systems must be resilient, meaning designed to cope with failure; the system stays responsive and working, even in the face of failures. Resilience is achieved by replication containment, isolation and delegation. Failures are confined within containers, isolating components from each other and thereby ensuring that parts of the system can fail and recover without compromising the system as a whole. Recovery of each component is delegated to another (external) component and high-availability is ensured by replication where necessary. The client of a component is not burdened with handling its failures (Reactive 2014).

Moreover, the system must stay responsive under varied workloads; as we have seen, this concept is called elasticity. Reactive Systems can react to changes in the workload by increasing or decreasing the resources allocated; this implies that the designs have no contention points or central bottlenecks, resulting in the ability to share or replicate components and distribute inputs among them. Reactive Systems support predictive, as well as Reactive, scaling algorithms by providing relevant live performance measurements. They also achieve elasticity in a cost-effective way on commodity hardware and software platforms.

In addition, Reactive Systems rely on asynchronous messages that establish a boundary between components; this boundary ensures loose coupling, isolation, location transparency, and provides the means to delegate errors as messages. Employing explicit message-passing enables load management, elasticity, and flow control by shaping and monitoring the message queues in the system and applying back-pressure when necessary. Location transparent messaging as a means of communication makes it possible for failure management to work with the same constructs and semantics across a cluster or within a single host. Non-blocking communication allows recipients to only consume resources while active, leading to less system overhead (Reactive 2014) and reduced costs.

Emerging high-performance applications require the ability to exploit diverse, geographically distributed resources based on Reactive Systems that can be implemented on a Cloud solution, maintaining the same level of self-provisioning and fast delivery of the traditional "shared-everything" configurations; in other words, they must be DevOps enabled. While the physical infrastructure to build such systems is becoming widespread, the heterogeneous and dynamic nature of the environment poses new challenges for developers of system software, parallel tools, and applications.

The introduction of these architectures, composed by different small, economic and replicable elements, interconnected by high-speed networks, on the public cloud and on premise, in fact generates new complexities in terms of configuration and system management. These complexities are the results of a need to create virtual environments in a simple and fast way and also to have the capacity to replicate them on a grand scale. To manage large installations of different elements generated and configured by the software (the so called Software Defined Infrastructure-SDI-), the virtualization concepts, already available in computing power, must also be feasible for other architectural elements such as the network, storage, firewall, and balancers.

The SDI is a concept for integrated control and management of converged heterogeneous resources enabling programmability of infrastructures. Together, cloud computing and Software-defined networking (SDN) promise a future open marketplace where applications can be readily and rapidly programmed on a converged infrastructure. Major collaborative open source efforts are helping to advance the following two approaches: OpenStack for cloud computing and OpenFlow for SDN (J-M Kang et al. 2014).

The requirement is coming from the new Cloud; this Cloud will be multi-tiered, with massive remote data centers in one tier as well as converged smart edge nodes in another tier, closer to the users. These nodes are essential for supporting applications with low-latency requirements, executing security functions, and promoting efficient content distribution through local caching resources. The remote, large datacenters leverage the computing virtualization and the networking resources to deliver flexibility and compelling economies of scale. In contrast, the smart edge is significantly smaller in scale with much more heterogeneous resources (J-M Kang et al. 2014).

Open interfaces for controlling and managing these shared heterogeneous resources can provide software programmability for dynamically deploying new functionality, as needed by a modern self-provisioned infrastructure and DevOps. In addition, advanced monitoring and measurement techniques and user access to infrastructure information can provide customized resource allocation or networking as required by the Reactive Systems elasticity.

The external entities can be applications, users (service developers or providers), and high-level management systems. The SDI manager typically performs coordinated and integrated resource management for converged heterogeneous resources through a resource controller and a topology manager. By adopting a "Software Defined Infrastructure" you can automatize (almost) everything from software lifecycle to monitoring and problem investigation. Taking advantage of the new architecture (Microservices as well as Elastic Resource Manager running Containerized systems/applications) renders it easy to build an effective automation process (part of DevOps) that strongly supports Development as well as Operations in implementing & delivering services faster and safer. Enhanced DevOps should not be considered optional: it allows you to run complex systems with lower costs and providing timely and effective services. Forget building a reactive system without it.

We just mentioned Containerized systems/applications; let us now come back to the interstate metaphor to better understand the concept of software Container and its importance. How difficult is it to move goods from one truck to another one? And what If we must also use ships, train, and airplanes during the journey?

Goods would be stored at warehouses until a vehicle is available. When an empty vehicle arrives, these goods would be moved from the warehouse typically using sacks, bales, crates and barrels, and then they would be loaded by hand onto the vehicle. As you can imagine this is a very labor intensive process and very expensive.

> On April 26, 1956, a crane lifted fifty-eight aluminum truck bodies aboard an ageing tanker ship moored in Newark, New Jersey. Five days later, the Ideal-X sailed into Houston, where fifty-eight trucks waited to take on the metal boxes and haul them to their destinations. Such was the beginning of a revolution. Decades later, when enormous trailer trucks rule the highways and trains hauling nothing but stacks of boxes rumble through the night, it is hard to fathom just how much the container has changed the world. In 1956, China was not the world's workshop. It was not routine for shoppers to find Brazilian shoes and Mexican vacuum cleaners in stores in the middle of Kansas. Japanese families did not eat beef from cattle raised in Wyoming, and French clothing designers did not have their exclusive apparel cut and sewn in Turkey or Vietnam. Before the container, transporting goods was expensive— so expensive that it did not pay to ship many things halfway across the country, much less halfway around the world. What is it about the container that is so important? Surely not the thing itself. A soulless aluminum or steel box held together with welds and rivets, with a wooden floor and two enormous doors at one end: the standard

container has all the romance of a tin can. The value of this utilitarian object lies not in what it is, but in how it is used. The container is at the core of a highly automated system for moving goods from anywhere, to anywhere, with a minimum of cost and complication on the way. The container made shipping cheap, and by doing so changed the shape of the world economy. (Levinson 2006).

What happened with shipping containers (Container 2015) is happening today with software containers, and with its most important reference platform: the Docker technology (Docker 2016a).

In order to establish the much-demanded software portability (Raj et al. 2015), we have been fiddling with virtualization techniques and tools for quite a long time. The inhibiting dependency factor between software and hardware needs to be eliminated using forms of abstraction so that any software can run on any hardware; for some time now, we have been able to do that by creating multiple virtual machines (VMs) out of a single physical server. Each VM has its own operating system (OS), isolated from the others; these VMs share the same physical hardware through automated tools and controlled resource sharing. This way, heterogeneous applications, based initially on different software and hardware architectures are able to run through a single physical machine, making IT infrastructures open, programmable, remotely monitorable, manageable, and maintainable (Raj et al. 2015).

Unfortunately, the virtualization paradigm has its own drawbacks:

- because of the VM carrying its own operating system, each VM provision typically takes a while, the performance decreases due to excessive computational resource usage, and so on;
- the growing need for portability dictated by SDI and Cloud is not fully met by virtualization. As a result, hypervisor software from different vendors was introduced to ensure application portability, but creating compatibility issues. "Differences in the OS and application distributions, versions, editions, and patches hinder smooth portability" (Raj et al. 2015);
- computer virtualization is common practice but SDI, with the concepts of network and storage virtualization, is just taking off;
- the building of distributed applications through VM interactions is complicated and sometimes prone to error;
- virtual machines need a few minutes to initiate, which can impact the user experience and give hackers time to exploit known vulnerabilities during bootstrap (CISCO 2014);
- patching and lifecycle management for virtual machines require a significant effort since virtualized applications have at least two operating systems for operators to manage and secure: the hypervisor and the guest OS inside the virtual machine (CISCO 2014);
- even the simplest OS process needs its own virtual machine. This requirement increases flexibility, but it also makes virtual machines impractical to use for providing microservices to architectures with hundreds or thousands of processes (CISCO 2014);

Designing the New Digital Innovation Environment   157

- when each physical server is replaced by one virtual machine, physical resource utilization tends to remain low. Server sprawl is simply replaced with virtual machine sprawl (CISCO 2014) and it is very rare to find a true capacity management on the VM farm above and beyond the automatic one.

All of these difficulties contribute to the unprecedented success of the idea of containerization. A container should contain only an application (or in the extreme concept of containerized micro-services, one micro-service); all of the application's dependencies (libraries, binaries, etc.) are bundled together as a comprehensive, isolated, portable, entity for the underlying infrastructure to be able to run it. Containers are exceptionally lightweight, highly portable, and easily and quickly built (Raj et al. 2015).

The DevOps goal gets fulfilled through application containers. The popular containerization platform, Docker, has come up with an enabling engine, downloadable as an open source, to simplify and accelerate the life cycle management of containers. Industry-strength Solutions are already available for container networking, management and orchestration, making the production and sustainability of business-critical distributed applications much easier. While Docker has made it easier to create and delete containers, the container community is currently developing new tools for continuous delivery and advanced testing of container images. Vendors are also working to create an audit trail for containers, adding metadata to the container images to show their content and when and where containers are delivered. Metadata might also include information about who produced the container, the container's products and components (for license management), and certifications (CISCO 2014).

Moreover, the cloud providers are working to soon offer a new set of container-centered services, creating a new stack called Container As A Service (CaaS); this incentivizes the containerized business workloads to move to cloud environments.

Precisely speaking, containers are turning out to be the most featured, favored, and fine-tuned runtime environment for IT and business services (Raj et al. 2015).

2.2 Applications

So, requirements and infrastructures have deeply changed and IT engineers need to act quickly under business and competition pressure. Whether you are moving to the cloud, migrating between clouds, attacking and modernizing legacy or developing new apps and data structures, the desired result is always the same: speed. "The faster you can move defines your success as a company"(Docker 2016b).

But how can we increase speed and quality while developing our applications? For certain, the development methodology is a key aspect, but the underlying architecture and the DevOps enablement are fundamental elements. In this search for new reference platforms, developers discovered that Platform as a Service

(PaaS) models were too high-level, abstracted and restrictive, requiring a lot of compromises about flexibility in favor of simplified operations and black box management. Similarly, Infrastructure as a Service (IaaS) is not sufficient as it provides a view of what resides solely in that infrastructure silo and it also requires the developer to have a strong background in operations to really exploit the environment. In addition, the more we move from IaaS to PaaS, the more the future portability of the solution is challenged. Organizations are therefore more so using a Containers as a Service (CaaS) environment to provide agility for development teams, control for operations teams and portability of apps across any infrastructure, from on-premises data centers to public cloud, across a vast array of network and storage providers (Docker 2016b). The Docker platform provides an integrated suite of capabilities to support the CaaS model; IT operations teams are able to secure, provision and manage both infrastructure resources and base app content, while developers are able to build and deploy their apps in a self- service manner (Docker 2016b).

All of this sounds great, but as always in the life, there is a price to pay. The price in this case is that, to fully take advantage of these infrastructural architectures, we have to rethink the way the applications are built. The elastic infrastructure requires a modular, distributed, autonomous node that can be duplicated on demand to scale up horizontally; the application should then be functionally distributed across multiple computing elements, each on a stand-alone process. On the other hand, the speed, of which we spoke earlier, does not allow for long wait times and bureaucratic release management processes, as it requires high quality deliveries ensured by automated testing and by the performance of a continuous delivery in production. Also, in this case, modularity is a must; this requirement increases the parallelism in development by multiple teams (such as the Hamlets we will see later on) and reduces the risk of always having to modify the entire application solution.

In short, we need an architecture that also uses microservices. This is an approach "to develop a single application as a suite of small services, each running in its own process and communicating with lightweight mechanisms, often an HTTP resource API. These services are built around business capabilities and are independently deployable by fully automated deployment machinery" (Fowler and Lewis 2014). Instead of building one application (monolithic architecture), developers build a suite of components, called microservices, which come together over the network using Rest APIs. Each component is written in the best programming language and data technology for the task, and each component can be deployed and scaled independently of the others (CISCO 2014).

Containers are better suited for microservices than virtual machines because they can start up and shut down more quickly. Applications that benefit most from a microservices architecture tend to be horizontally scalable. To take advantage of the containers scalability (computing, memory, and other resources), vendors work to create frameworks for managing container images and orchestrating the container lifecycle, in accordance with the workload. The Reactive Manifesto, with its elasticity, resiliency and responsiveness, can, in this way, become a reality also for large traditional transactional Information Systems. An application based on microservices has clearly defined boundaries and dependencies, and is highly flexible for operations, maintenance and future changes. In fact, each microservice is a separate entity and is deployed as an isolated service; it can change or scale separately from other services. Being small in size, the price of replacing them with a better implementation is quite affordable, reducing the overall functional depth.

But microservices and containers alone cannot perform the miracle of building a modern Information System. There are a number of characteristics that a "new normal" software solution should have. Due to space constraints, it is impossible to do them all justice here, but the intent is to create a synthetic take-away that could be used as a sort of check-list.

Here it is, not in order of importance:

- **Be CaaS ready**: containers allow for the packaging of a service with all of its dependencies into a standardized unit of deployment that can be executed by a distributed resource manager, and provides: (a) Isolation between different technologies stacks (b) Portability across different physical infrastructures, even to Cloud (c) Scalability and reliability (HA & DR). An architecture based on containers and on a unique distributed resource manager will allow efficiency while supporting flexibility both in the short and in the long term (adaptable to future technologies/future scenarios). Work on premise to get around temporary "public" issues and start using the public for development environments.
- **Containers**: applications must be able to run under containers and to be orchestrated (horizontal distributed workload management).
- **Shared nothing**: the applications must be able to run on a Shared Nothing Architecture. This is required for properly exploiting advantages of a cloud ready, distributed environment, for instance fast provisioning/decommissioning of machines, standardization of hardware and automatic workload balancing. The perfect environment for experimenting with the CaaS is the application development environment, meaning all the infrastructures and tools needed to develop a solution.
- **Single Page Application**: modern applications require dynamic web applications with real-time updates without page refresh; scalability and high performance together with very rich interfaces are absolute requirements for web development today. The answer to this demand is a new way to develop web content, called Single Page Web Application (SPA). This kind of development allows you to write less server-side code and more client-side code, which is sometimes focused on JavaScript, providing a better user experience with a new way to interact with the application. Products such as Gmail, Trello, and Groupon are examples of successful SPA development (Monteiro 2014). Along with the new development approach, a new set of frameworks has been delivered to ease the developer's life, unleashing the capabilities of a new generation of interface model like "the material design" of Google. Examples of these frameworks are JQuery, AngularJS and AngularJS 2.

- **Be Composable**: large systems are composed of smaller ones and therefore depend on the reactive properties of their constituents. This means that Reactive Systems apply design principles, so these properties apply at all levels of scale, making them composable. The largest systems in the world rely on architectures based on these properties and serve the needs of billions of people daily. It is mandatory to apply these design principles consciously from the start instead of constantly having to rediscover them.
- **Be designed by Patterns and Data Driven**: the need for autonomy and unleashing of creativity must reach a compromise with common good practices and a common way of doing the same thing without analyzing and solving the same problem (may be in different ways) more than once. An example of such a compromise is designing architectural patterns for modernization and new solutions. Designers will be able to count on a library of patterns that they can use or extend as needed. Another important aspect regards design decisions, very often about the User Experience. Usually, a person or a team makes the decision, with more or less appropriated information; to be under "Data Driven Decision" means that the identified problem and the different options are measured in production using a small number of selected customers. It also means that the project leader will consider that data for the decision. If one has to decide if a button should be put on the right top or on the right bottom, for example, both options should be tried with a different set of users to measure the impact it has on them; in this way, the definitive position is decided based on data and not on the gut feeling of a manager (Kniber 2014).
- **Be Resilient by nature**: "Chaos Monkey"is for sure an extreme concept about resiliency. In truly fault-tolerant systems, it could make sense to increase the rate of faults by triggering them deliberately. Software that deliberately causes faults —for example, randomly killing individual processes without warning—is known as Chaos Monkey. This ensures that the fault-tolerance mechanism is continually exercised and tested, so that we can be confident that faults will be handled correctly when they occur naturally. Even if we do not want to use the Chaos Monkey, we need our solution to be truly resilient and the redundancy to manage possible faults automatically and seamlessly.
- **Be Sustainable**: focus on sustainability should be a mantra. We should develop only those functions which make a difference in the Customer/Company experience. It should be always the goal to obtain the "Minimum Valuable Product" in all the designs of new solutions.
- **Be Productive**: the developers are under pressure due to increases in demand and speed. The time to market is becoming an obsession, quality requirements are constantly becoming more stringent, skills are sometimes scarce, there is pressure to increase capacity or to do more with less. We need to use the same approach that Open Source developers are using to create their artifacts; even though they are working from different locations, they produce high quality and innovative code through a special collaborative development.

- **Be Open**: openness is a key aspect. Being open is a very important feature of future development approaches. One must understand that information systems are evolving toward a multitude of microservices federated together that will collaborate with each other, even when distributed across multiple different infrastructure providers. Those who consume these services will often be unknown to those who provide them, as well as the workload will not be known. The interfaces between the different services must be described and standardized; in other words, even if the interface is open to external users, it would be better to document and manage it in the API Gateway. In this way, you increase the isolation of the different components of your Information System and increase the possibility of development parallelism across different teams. The Open Source context is also a source of inspiration for innovation; a lot of new technologies and new methodologies emerged from open source projects.
- **Be Social**: collaboration and social interactions should be played up. The environment should be very inclusive, providing tools that easily integrate the different roles into a single body. This is what we call "Social Development Environment" and in short it should contain functions to facilitate the following activities:
    - knowledge management—embedded search on every object treated; code sharing; best practice sharing; automated feeding from the day-by-day activities; automatic documentation of processes and solutions;
    - collaboration—networked task management; easy cross communication; teams' material integration; net-organization; communities of practice;
    - productivity—DevOps connection; automatic Wiki creation; GIT support; build and automatic testing support.
- **Enable Continuous Delivery**: Continuous delivery is a set of tools and processes that allow code to be rapidly and safely deployed to production, ensuring the quality of business applications through rigorous automated testing. This means that every change is proven to be deployable at any time. The micro-services and containers facilitate the adoption of this approach.
- **Enable automatic testing**: it should be seen as complementary of Continuous Delivery. Despite the several tools now available to perform automatic tests as well as the new design methodologies to define test cases from the beginning of projects (e.g. Test Driven Design), automatic testing is not a normal practice in most companies. The approach also influences change and release management processes; it should not be gated by a never-ending number of procedural checks and manual operations, but by a progressive introduction in the production environment using a targeted set of users, directly measuring the service quality. It is a data driven release management that, together with automatic testing and elastic infrastructure, can give a robust and high quality service while achieving a continuous delivery of changes.
- **Enable Data Centricity**: this means Data Integration and Analytics by design. The continuously growing demand for data availability and analysis match badly with traditional approaches (ETL/data federation/access to production

DB). Integration among data producers, journals and consumers must allow them to work together seamlessly and efficiently. Traditional applications are typically based on "processes"; the application resembles the different steps belonging to the process, assigning a role and responsibility to each step. The data associated with the step constitute a dependent variable and may be very specific to that step and that process. With this approach, data model harmonization and data integration can be a nightmare, with continuous data transformation and great difficulties for a user looking to navigate seamlessly through different activities (steps of different processes) without having to re-enter data. Since a modern user experience requires an easy navigation through application domains, an organization seeing data as the center instead of the process can dramatically change the customer experience. Moreover, the Big Data capabilities are encouraging common data models across information systems to tie together different data spaces and ecosystems.

- **Support Choreography**: it is a different approach with respect to Orchestration. It implements a publish-subscribe paradigm. It provides and guarantees a higher degree of decoupling between services, and therefore a better flexibility and maintainability. It re-defines transaction boundaries and asynchronous services allowing for an "Event Driven" company.
- **Support Journaling**: a Journal stores ordered sequence of events. By using the journal as a master copy for your data (the history of changes), you enable the creation of new architectural patterns, aimed at reducing the gap between operational databases and analytical systems: i.e. near real time updates on both systems, replay/playback for handling errors, retry or compensations (time windowing applications). It allows for a flexible interface between data producers (Data Systems, microservices, Events) and Data consumers. It could be very effective in a data centric architecture.
- **Avoid "One-size-does-fit-all"**: a unique standard cannot fit all requirements. We have to define lists of standards targeted on specific use cases, leveraging opportunities provided by highly specialized tools (no more standards for product types/product families). It is mandatory to use containers and microservices to avoid cross-dependencies. If we want to use the Reactive, Shared Nothing Architecture, the standardization of the different layers and stacks is unavoidable (especially to use DevOps and Cloud), but defining specific non-normalized stacks within a container should be permitted. In other words, we should not care about what is in the container, and instead give the development teams the highest flexibility and freedom to select what works best. We should, however, be integralist on the CaaS architecture and on the fact that the container should be runnable on any infrastructure and tested to be cloud ready.
- **Avoid "everything is a transaction"**: transactions are the backbone of our systems. When moving toward a distributed shared nothing architecture (maybe on Cloud), the handling of distributed transactions across different servers is expensive both in terms of resources and performances. New generations of data systems often redefined the word transaction to describe a much weaker set of guarantees than databases, or even, entirely abandoned transactions in the name

of scalability, availability and performance. This redefining of transaction boundaries in accordance with business functions' intrinsic architecture is a necessity and an opportunity for building loosely coupled services, and thusly gaining in flexibility and maintainability. Sometimes the abuse of transactional patterns may jeopardize the simplicity of an Information System and make the adoption of microservices very difficult.
- **Avoid "everything is a Database"**: we have to consider new technologies for handling data, optimized for a variety of different use cases (document stores, key-values stores, full-text search engine, graph databases, journaling etc.) and move from Databases to Data Systems. Also, approaches like Data As A Service may facilitate the adoption of different information management strategies.

2.3 Information

We, as human beings, have been generating data for thousands of years (Mohanty et al. 2013). The different technologies that humans have invented over the years facilitated the creation and diffusion of data, but nothing boosted the information production as much as Information Technology; what we have seen so far is nothing compare with what has yet to happen. Every two days, we create as much as information as we did from the dawn of civilization up until 2003, around 5 exabytes of data (Schmid 2010), that is 2,5 exabytes per day (Price 2015), with 90% of all the world data being produced within the last two years (Price 2015); the estimated total size of this data by 2020 is 44 zettabytes, up from only 4.4 zettabytes in 2013 (IDC 2014). But what do we do with this data and how is it used? For years an overwhelming amount of data was deemed useless (Mohanty et al. 2013), mostly that which was not in digital format and stored without any order, making very difficult to process, retrieve and analyze. Still this data was important for every enterprise, regardless of its size and, over the years, we learned how to structure the data, how to govern it, and how to consume it in a faster and more efficient way. These developments were possible through the creation of new approaches and methods to govern data within firewall boundaries; the order of magnitude was a few terabytes.

But it was with "user-generated content" or prosumerization (the simultaneous production and consumption of data) that the amount of unstructured and interactive data assumed the shape that we all know today, a veritable data explosion. It was with the dominium of the free search and "I feel lucky" paradigm that we entered a new age of data navigation, without worrying about the incredible and ever-growing amount of data, but rather being amazed by the business opportunities that this phenomenon may generate.

Another emerging perspective concerns how the enterprise boundaries are managed; in the past, the majority of data was within the corporate firewalls, (internal data) while today, if you want to understand the influence of your company on the full market, it is almost the opposite (external data). A new bison's paradigm of "consumerism" was born (Mohanty et al. 2013).

The famous and inflated term Big Data basically tries to describe three characteristics and a challenge. The three characteristics are identified with the 3Vs: Large data **V**olumes, the high **V**elocity at which the data is generated, and the **V**ariety offered by the data. The challenge is represented by the technical and governance aspects of managing such types of data, without forgetting the huge amount of internal data that was impossible to mine in the past due to the size, time of processing and costs.

Around new kinds of data, new and fascinating functions were made available (Mohanty et al. 2013; Akerkar 2014):

- Search with fuzzy matching, Content-Based Image Retrieval (CBIR), Natural Language Processing, clustering, sophisticated automatic filters
- Multimedia content
- Sentiment Analysis, including brand and reputation analysis
- Enriching and contextualizing data with the purpose of having a better quality of master data management, also using interfaces like OAUTH2 for common sign-on and user authentication
- Real-Time Big Data Processing for Domain Experts, such as Smart Buildings, wearables, start home, sensors and Internet of Everything in general
- Data Mining and Discovery
- Operational analytics
- Machine Learning, Deep Learning and Cognitive computing
- Knowledge creation.

These functions need new architectures and engineering paradigms; the first massive use of data within the enterprise was based on the "Collect-Process-Manage-Generate" process (Krishnan 2013), with the collected data that is structured in nature and the entire process based on known requirements. Nevertheless, governing this small and well known amount of data was difficult; instrumentation of data required a complete understanding of the data to maintain consistency through the entire processing cycle, integrating multiple and different data sets. The amount and kinds of data still allowed RDMS and OLTP to handle the situation, but there are at least four different areas representing a challenge:

- **Storage**: storage has historically been the first problem to become visible (Krishnan 2013). The burn rate of the storage increased dramatically in the last decades. The evolution in technology and the decrease of the cost per gigabyte were not enough to absorb the cost increase due to the demand of storage capacity. The storage life cycle and the different tiers were not often applied, resulting in issues of performance and total cost of ownership. Moreover, it boosted governance complexity in terms of operations, architecture and integrity. Other aspects, such as Business Continuity and Security, were often neglected.

- **Transportation**: Another big issue is moving data between different systems and then storing it or loading it into memory for manipulation (Krishnan 2013). Even if networks have evolved considerably in the years, in terms of connection speed and bandwidth, they still bottleneck and represent one of the major reasons to rethink the architecture in different terms. Also, the different data domains have created the need for moving and transforming the data to be coherent with the different types of use.
- **Processing**: CPU, Memory and software combine the logic and mathematical computation required for data processing (Krishnan 2013). The capabilities of these elements have been improving, but, nevertheless, they still represent a challenge, especially for low cost and large scale computation. The new architecture should be able to exploit the availability of a large amount of low cost equipment federated together to produce a high throughput.
- **Integration and Data Quality**: Always more regulators and auditors have begun requiring that every manipulation of data be tracked and kept under control. We are still far from a full endorsement of the Single-Point-of-Truth, but the request of a consistent mapping of all the data entities, their manipulation, the security access, the privacy implications, and the description of all uses are becoming increasingly more urgent. This is especially true when different data sources are mixed-in and heavy data transformation is involved.

All the above challenges have created, over time, problems in terms of speed and/or throughput of data processing. Speed is the combination of various architecture layers: processors, memory, software, networking and storage. Each element has its own limitations, but the combination of all of them has been, in some cases, detrimental to data processing capabilities (Krishnan 2013).

These types of architecture can generally be placed under the Shared-Everything category described above. The Shared-Everything solutions are typically used for transactional applications demanding high performance; in recent times, however, the shared-nothing architecture has also been suggested for high transactional workloads. But how to manage the data consistency and integrity in a shared nothing architecture is open to discussion, driving the use of a new generation of DBMS that are suitable for both structured and not structured data, and able to perform particularly well under generic searches and queries.

In fact, in traditional data processing, like Data Warehouse, you try to understand the data by creating a set of requirements that drive data discovery and data modelling; in the end, a database structure is created to process the data. Obviously, this approach is optimized to manage the data in the end state, favoring write performances. In Big Data processing, the data is collected and loaded onto a target platform, metadata are applied and a data structure is created for the content, to transform and analyze the data (Krishnan 2013); due to the volumes involved, a file-driven architecture with a programming language is preferable to a database driven architecture. The requirements for a Big Data architecture can therefore be summarized as follows:

- No data-model (or almost)
- Near real time data collection
- Micro-batch processing
- Minimal data transformation
- Efficient data reads
- Store results in file system or non-relational database (Krishnan 2013)
- Programming language interface
- High performance
- 99,999 (5 nines) availability
- Distributed processing with horizontal scalability (including shared data across different nodes).

This architecture is very close to what is needed for a modern Web application; it therefore constitutes a concrete opportunity to redesign solutions so that they reach the objective of a shared nothing architecture, that is, using an elastic infrastructure while also putting data at the center. Such a change will permit the application layer to see continued growth over the next few years in an attempt to make Big Data accessible to an ever broader audience (Feinleib 2014).

Once you endorse the right architecture for building Big Data applications, you have to give value to the data. The analysis of the data and the tools that are available to us change over time. In particular, these tools are becoming more available (anytime, anywhere), easy to use and graphically appealing. Such features are typical of a mobile-cloud world in which you are fed by a growing number of applications for both business and everyday life. Fitness, healthcare, sales, and logistics are just a few examples of situations in which the mobile, cloud and Big Data are joining together to produce new and intriguing applications. Big Data is changing the mobile and the mobile is changing Big Data (Feinleib 2014).

It is also worth mentioning the importance of machine learning and deep learning, as they have a significant impact on the exploitation of Big Data and future opportunities connected with the High-Powered Peripherals.

Somebody thinks that large investments in developing and researching new algorithms could be the cause of technological singularity. A technological singularity is a hypothetical event where an intelligent upgradeable agent (such as an artificial intelligence-based software) enters a 'runaway reaction' of self-improvement cycles, with each new generation more quick and intelligent than the last, thusly creating a "super-intelligence" whose cognitive skills may be qualitatively much superior to those of humans. According to Raymond Kurzweil, the predicted year for this singularity is 2045.

More generally, the term has historically been used for any form of exponential acceleration of technological progress that can potentially cause discontinuity, in which events become unpredictable or even incomprehensible to human intelligence. Above and beyond predictions, we can improve our Information Systems and Enterprises, using these new approaches in several areas, such as the recognition of texts, images, voices, and videos, or the understanding of natural language, with the ability to answer questions. The applicability is in numerous sectors such as, banking, insurance, transport, retail, internet or telecommunications.

2.4 High-Powered Peripherals

Internet of Everything (IoE), Augmented Reality, and Wearables influence each other, creating a series of applications that, in combination with cloud and new mobile platforms, enable solutions that were simply unthinkable a few years ago. These applications have certainly had disruptive effects on our lives and on the way we interact with business and the economy in general.

As we have seen, Big Data is competing in this exponential transformation; we can monitor our heart rate, sleep patterns and even your EKG using mobile devices or accessories and accessible applications, finally uploading and sharing all the collected data using the cloud (Feinleib 2014).

One example speaks for all: Waze. This is a GPS app for avoiding traffic, thanks to probably the largest car driver community in the world.

Just by driving with Waze, you are already able to generate traffic information and road conditions in real time, helping the community of drivers in your area. You can report accidents, hazards, checkpoints and other events that you see on the street and also receive information on what is happening along your route. You can find the cheapest petrol station on your route thanks to other Wazers sharing gas prices. Also, it's fun and easy to meet friends, and schedule appointments. Waze is all a matter of "giving" and "sharing" to "draw" a common benefit while you are driving and it is melting together all the elements (mobile, sensors, cloud, big data) that we presented earlier.

Another example is Strava, which tracks your runs and rides with GPS; you can join challenges and see how your running and riding compare with friends. You can follow routes you've created or found and view your activity map as you record key stats like distance, pace, speed, elevation gained and calories burned. You can also collect heart rate, power and cadence data from Bluetooth LE sensors. Furthermore, you can find your friends and motivate them with kudos and comments so that the riding experience is social, even if the counterparts are not there with you at the same time.

A deep instrumentation is also happening in our cars, not to simply just record data on all the working parameters of our trips (drive style, consumptions, tires, etc.), but also to share this data to external applications that, in turn can federate together to produce additional services and value.

Our interaction with mobile devices is transforming, amplified by the fact that mobile phones are always on our person. It is the best example of the emerging need for "Cognitive Prostheses," which according to the Oxford Dictionary is "An electronic computational device that extends the capability of human cognition or sense perception."

This emerging concept of human-centered computing represents a significant shift in how we think about intelligent machines. The concept embodies a vision in which "human thought and action and technological systems are seen as inextricably linked and equally important aspects of analysis, design, and evaluation. The prosthesis metaphor implies the importance of designing systems that fit the human and machine components together in ways that synergistically exploit their respective strengths" (Ford 2001).

This approach focuses more on computational aids designed to amplify human cognitive and perceptual abilities, leveraging and extending human intellectual capacities. What is really important to note is that the need to always feel socially connected, to have all our data analysis shown in context in near real-time, to mix different sensors' data and combine that with historical trends or with what others are producing, is pushing us towards a full immersion into an augmented reality that is not yet available. All the attempts that we have seen so far (such as Google Glass, Microsoft HoloLens, Instagram Sunglasses) have worked towards fulfilling the prosthesis needs; even if the experience is not yet completed, it eventually will be, and by then, we will need a new generation of applications that will scale elastically and embed all the big data principles for exploiting machine-learning and deep-learning algorithms in near real-time. The design and fit of these new applications will require a broader interdisciplinary range of expertise including computer scientists, cognitive scientists, physicians, and social scientists. This necessary collaboration represents a challenge in addition to the challenges with methodologies, tools and skill availability.

The Cognitive Prosthesis and the multitude of connected applications and devices will result in a push for common digital identities for people, animals and objects that are recognizable, immediately and seamlessly, by the underlying platforms. This goal of this push is to achieve the expected federation of services and data exchange. This federation will increase the ability to track and trace everything, amplifying the risk of misusing data from a privacy point of view.

Above and beyond the Cognitive Prostheses, it is clear that IoT is a major phenomenon that cannot be ignored; its disruptiveness is intrinsic into its definition: "seamless combination of embedded intelligence, ubiquitous connectivity, and deep analytical insights that creates unique and disruptive value for companies, individuals, and societies" (Banter and Holman 2016). There will be over 30B connected devices by 2020 and IoT will have an impact on around 6% of the global economy with the possibility of exponential growth.

There will be challenges and pressures that will determine winners and losers:

- Lack of platform standardization, lack of device standardization
- Lack of integration from an end-to-end services point of view, with the risk of insufficient support for new services
- Difficult interoperability with lack of integration from an applications point of view
- Difficulties in finding the right business cases to justify an early investment
- Policy makers and regulators not keeping up with the pace of innovation

Designing the New Digital Innovation Environment

- Exponentially growing volumes
- Difficulties in finding and creating the right skills
- Initial lack of scale to achieve the right marginal cost
- Too many alliances and consortiums
- Risk of newly created legacy to manage due to the rapid obsolescence of the solution and the lack of interoperability standards.

In other words, it will be stormy waters to navigate, but it will be impossible for any enterprise or individual to stay at the window waiting for good weather.

Regarding 3D Printing, we should use a different logic.

This important technology is working somehow more independently than other High Powered Peripherals; therefore, it seems to not concur with the combined effect of mutual acceleration. Nevertheless, 3D printing technologies are actively tested and used in a broad spectrum of industries, from health care to aerospace. There are noteworthy cases of 3D printing for mass production (e.g. hearing aids, dental braces, LEAP engine nozzles), but the dominant reason for their existence is to accelerate manufacturing time-to-market by prototyping artefacts with this technology; as 3-D printers become faster, easier to use, able to handle multiple materials, and print active components or systems, they will definitively find use beyond rapid prototyping (Earls and Baya 2014).

We refer to 3D printing typically as a unique technology; in reality, the many different printing technologies are generally material dependent. For instance, fused filament fabrication (FFF) is used with plastics, stereolithography with photosensitive polymers, laser sintering with metals, while laminated object manufacturing (LOM) bonds laminates of materials (such as metals, plastics, or paper) in successive layers, and so on (Earls and Baya 2014).

This fragmentation in different types of sub-technologies is not helping the improvement of the factors that are somehow slowing down the full function adoption of 3D printing.

The following improvements should be considered to facilitate 3D printing:

- Improve key performance characteristics (such as speed, resolution, autonomous operation, ease of use, reliability, and repeatability).
- Incorporate multiple types of materials and printing technologies (including the ability to mix materials while printing a single object).
- Provide the ability to print fully functional and active systems that incorporate many modules (such as embedded sensors, batteries, electronics, micro-electromechanical systems, and others).
- Lower the cost of mass production.
- Reduce the labor requirements associated with the printer by making it easier to install, maintain and operate.

The 3D printing industry is listening to these needs; experts predict that 3D printing will become 50% cheaper and up to 400% faster in the next five years (Siemens, Columbus 2015).

Despite these limitations, 3D Printing is continuing its rapid growth with a CAGR of 20,6% between 2013 and 2020 (Columbus 2015) and is expected to revolutionize industry value chains overtime. In fact, the innovation in 3D printing hardware has already reached larger artefacts, greater precision and finer resolution at higher speeds and lower costs; together, these advances have brought the technology to a tipping point (Cohen et al. 2014), ready to emerge from its niche status and be used in a larger number of applications.

CSC, by 2012, had already created a roadmap of 3D printing adoption across several industries describing what they can expect immediately and in the future (Koff and Gustafson 2012). These industries are: Defense & Aerospace, Automotive, Healthcare, Consumer & Retail, General Manufacturing, Supply Chain and Commercial; it is enough to give a glance at this roadmap to realize the magnitude of innovation and transformation that these ecosystems will require. The technology will transform manufacturing flexibility while making it possible to create complex shapes and structures that weren't feasible before. Moreover, additive manufacturing will help companies improve the productivity of materials by eliminating the waste that distresses traditional (subtractive) manufacturing, thusly creating value from a circular economy. The exponential economic impacts of 3D printing are estimated by the McKinsey Global Institute research as being up to $550 billion a year by 2025 (Cohen et al. 2014).

It is also obvious that the Information Systems built to support these ecosystems will be strongly affected; we suppose that the architectures working for the exponential businesses will also be able to support the transformation induced by 3D printing.

3 New Ways to Interpret the Software Development

It is clear that many new technologies will be part of new and more pervasive IT solutions, and that we cannot continue to manage IT as we have so far.

As you will find in other sections of the book, the current thinking is that there is not a unique way to manage the IT of a large enterprise and that there are at least two action speeds that we should respect, using two different interpretations of software development processes. The first approach, **Foundational**, is oriented around the foundations or the core of the IT assets and suited for well-known requirements and solutions; the value in this case will be around efficiency and price per performance while the development cycles look to long-medium terms. The other approach, **Adaptive**, should be used to deal with uncertainty or areas where we need to experiment; the value, in this case, is in the revenue, return on Brand, new products, and a better customer experience; the rhythm is much faster and is expressed in months, weeks, or even days.

To be competitive and to be sustainable in a large enterprise, we must also take care of two other ways of interpreting the IT activities pushed by the exponential businesses and the need for digital transformation: the Traditional IT and the Transformational IT. Without giving too many details about these two interpretations, we can say that the Traditional IT represents a typical automation approach where quality of delivery and efficiency are the key success indicators, and IT is not meant as the core business. The Transformational way is more exploratory, with a strong focus on IT as a lever for competition and business advantage; in this case, IT is often part of the core business of the company.

While the differences in terms, processes, roles and tools are noticeable between Foundational-Adaptive, those between Traditional-Transformational are more radical, citing different skills, different types of people, different organizational models, and of course a different operational model.

Because of these four dimensions, the person in charge of the IT Governance of the Enterprise will have the complex task of providing guidance on how those four approaches should be used and when. For each one, this person must define proper processes and tools. It may also be very difficult to re-skill/acquire/assign the required personnel over time, based on the different needs of the industrial plans.

To mitigate these difficulties, we can identify a few basic principles applicable regardless of which approach is selected; these principles allow for consistency at the enterprise level. The teams can also develop extensions to these principles to better match their mission and objectives. These principles are aimed at generating a new development culture and fostering innovation and evolution.

Before introducing these software development principles, it is important to have a full picture of the changes that have been happening in this context starting with a few reflections.

3.1 Reflection 1: The Assets Syndrome and Hierarchy

Humans have always been building goods and then selling access to them. This behavior, first seen in tribes and then adopted by clans, has spread to all nations, empires and most recently to the global markets, making larger human aggregations and institutions possible. The creation of value was based on the possession of large amounts of land, resources, machinery, or labor force. Owning was the perfect strategy to deal with a shortage of resources and to ensure a relatively predictable and stable environment (Ismail 2015, p. 50).

Once an organization had all the people necessary for managing and protecting assets, hierarchical organizations were created: in each tribe there was a hierarchical order, implicit or explicit, connected to a structure, for the purpose of managing power. Later, starting in the Middle Ages, even if only actively with the industrial revolution and modern businesses, the local and hierarchical mindset was reproduced in enterprises and governmental structures: this model has survived even today in most cases (Ismail 2015, p. 51).

3.2 Reflection 2: The Organizational Anomie

We just saw that hierarchical order is somehow derived from the need to manage assets on a large scale. The same need applies when we consider knowledge an asset. Overtime, we have observed a continuous increase in the depth of knowledge, with vertical specializations in increasingly narrow fields of applications. The combination of very specialized domains of knowledge and the presence of complex problems, demonstrates that single individuals could not have all the required skills, but rather need to rely on a set of people to run the task: the team.

Unfortunately, the underlying organizational models remain the same; the hierarchical model used to keep assets under management control remain unchanged in spite of the deeper specialization, the smaller size of artefacts, and the higher number of people required to perform a task. The "task force" concept emerged as a way to cope with this new tension between a need for horizontal co-teaming and respect for the hierarchy; the task force represents a temporary suspension of the hierarchy in the name of greater interests, such as responding to an emergency, an issue, or completing a project by the deadline. Matrix organizations, in some cases characterized by multiple dimensions, are an attempt to include the task force in a sort of predefined and structured organizational schema. This model, frequently even effective and efficient for the business, is often unintuitive and outside of our nature of living in simple social organizations with simple points of reference. In the end, typical organization models do not seem to be up with the times and create anomies, which Robert K. Merton defines as: a lack of balance, even due to the presence of obstacles, between existential purposes provided by the social culture and legitimate means available to achieve them. In other words, the companies sometimes ask employees to reach objectives, while setting processes and rules that represent artificial obstacles for their job, instead of enabling them and facilitating their job with appropriate processes and rules. Often, this anomie is the source of individuals refusing to follow rules in which they don't believe, and generates the search for a greater decisional autonomy, especially in software development.

3.3 Reflection 3: Linearity Vs Exponential

On the other hand, another pressure is pushing for changes: today we continue to measure our results on a linear scale (Ismail 2015, p. 51) while we are often in an exponential business, or product, or the organization is engaged in a survival game (in which you either destroy the competitor or you get destroyed). Put simply, we are used to seeing that a quantity of work (x) requires an amount of resources (y), 2x requires 2y and so on; the automation, mass production, robotics and virtualization through computers have changed the slope of this line, though it still remains linear (Ismail 2015).

In business, the construction of most products and services continues to reflect this linear, incremental, and sequential thinking. The classical method for the development of a product, from a big aircraft to a small microprocessor, is thought to be a standard stage-gate procedure called *New Product Development.* In the software development world, this procedure has been replaced by the waterfall approach (Ismail 2015); although the waterfall process is not the best fit for responding to current challenges, many people and organizations continue to try to make it work (Schwaber and Sutherland 2012).

Here are few examples of typical problems (Kelly 2008) not necessarily associated with the stage-gate approach:

- Late delivery.
- Over budget.
- Too many defects/bugs.
- Lack of functionality.
- Represents marginal improvement; for example, only automates existing practices.
- Does not meet the business needs; for example, fails to address the core problem.
- Too many features.
- Too slow.
- Does not match the competitor's product.
- The competitor was first to the market.
- Poor usability; too difficult to use, so employees avoid using it.
- Does not justify the expenditure.
- Too costly to support.
- It limits future changes.
- The company has changed direction, been taken over, or exited the business.

3.4 Reflection 4: The Engineering Mindset

To complete the context and frame the requirements for a new software development we also have to add another element: the engineering mindset with which we create solutions.

In the early days, most software development was a chaotic activity, in many cases under the label of "code and fix". Software was written like an artistic artefact, with little planning and many tactical decisions. This may work well when the system is limited in size, but as systems grow, it becomes increasingly difficult to maintain and develop solutions. Behind an apparent flexibility and speed, it becomes more and more difficult to keep up quality (Fowler 2005). From this situation, "Methodologies" arose, inspired by such disciplines as civil and mechanical engineering. These disciplines put a lot of emphasis on planning before building as well as working on a series of designs that pitch exactly what is needed to be put together. As such, the methodologies impose disciplined processes with the aim of making development more efficient and more predictable, precisely with a strong focus on the planning side. Ultimately, designs are passed to different groups or companies for execution (Fowler 2005). It is also worth noting that many past assumptions regarding software development ended up not always being true:

(a) you know exactly what the user requirements are; (b) you know all the technical difficulties up front and any problems will be small and manageable; (c) between when you start and when you finish, the tasks and the context will remain mostly the same; (d) the implementers will do exactly what is defined by the design; (e) the design and implementation activities can be well separated and performed by different people/companies; (f) you can reduce the required skill level by dividing the activities into small, distinct, autonomous tasks.

And of course, the model typically followed the New Product Development approach based on subsequent stages.

3.5 Reflection 5: The Bureaucracy

Keeping this reasoning general and at high level, *"the most frequent criticism of these methodologies is that they are bureaucratic. There's so much stuff to do to follow the methodology that the whole pace of development slows down"* (Fowler 2005).

Agile methodologies have been developed as a reaction to the bureaucracy of the engineering methodologies and the frequent failures of large software projects. These new methodologies try to reach a balance between no process and too much process, and to be less document-oriented, believing that the key part of documentation is the source code.

However, as Fowler describes, much deeper differences exist between the two methodologies:

- Agile methods are adaptive rather than predictive. Engineering methods try to plan everything, which works well until things change; therefore, they tend, by nature, to resist change. The agile methods, instead, welcome changes. "They try to be processes that adapt and thrive on changes, even to the point of changing themselves" (Fowler 2005).
- Agile methods are people-oriented rather than process-oriented. Engineering methods are designed not to be dependent on individuals. Agile methods are the opposite: the skill of the team can never be substituted by a process. The process should only be a scaffold and a support for the team.
- The Agile paradigm fights the hierarchical organization model in favor of a team based model (not very far from the Holocracy concept defined by A. Koestler in 1967) to be able to reduce the bureaucracy and the documentation to only the

essential. In this way, the teams are supposed to only spend time producing real things. The teams must also be very well skilled and small in size; less than eight members seems to be ideal.

In addition, evolutions are occurring on the technological side that could impact the way we develop our software, like containerization (e.g. Docker), micro-services, and continuous delivery. These evolutions will be detailed later.

3.6 Reflection 6: Enlarge Agile Principles

But how can we leverage these reflections to build principles that can help the agile methodology to be resistant in large enterprises, with huge projects, different skill sets and employees spread cross-territory and nations? Can the agile methodology face the challenges of digitalization and the exponential businesses?

To answer to these questions we first need to recall that Agile is not just a collection of software development techniques but a movement, a way of thinking, a culture; before we get any further in the description of the required principles, it is useful to look at the fundamental underlying assumptions, which agile proponents call "values", and how we can extend them to take care of all the inputs that we have seen earlier. The original values are (Meyer 2014, p. 3–4):

- Redefined roles for developers, managers and customers.
- No "Big Upfront" steps.
- Iterative development.
- Limited, negotiated functionality.
- Focus on quality, as achieved through testing.

Sometimes, in large enterprises, IT departments resist new approaches, including attempts at implementing the Agile methodology, preventing its massive and wide spread utilization.

To fulfil the requirements for modern software development in a large enterprise, and to modify the Agile methodology so that it is more acceptable to IT departments, we should enhance it with the following principles:

- Enforce a Trust culture
- Redefine teams and their interactions
- Redefine concepts behind standards
- Transform documentation in knowledge Management
- Embed architectural and operational key facts in the development process.

Let's try to briefly introduce each of them.

3.7 Augmenting the Agile Approach with New Principles

Enforce a Trust culture

Even though this is already part of the Agile culture, to win today's challenges, it is important to tap into the talent of everyone in the organization. The combination of a culture of trust and everybody knowing and owning results is the foundation for innovation and motivation (Pixton et al. 2014). In this context of continuous transformation and change, trusting each other is critical: it is almost impossible for a person "to know it all when it is all different" (Pixton et al. 2014). To really be more effective, we need to be able to count on all individuals who contribute. But trust and ownership alone are not enough, there must be also a passion for delivering the right results as part of the culture. In addition, to manage the uncertainty, we cannot increase control over the tasks; managers need to put the focus on "why something is done," working with the team on the ambiguity, in a transparent way. Last but not least, we should not blindly trust what has worked in the past, as it might no longer work (Pixton et al. 2014). The autonomy that has always fueled the Agile approach also has to conform to certain common themes, such as general directions of the company, reference architectures, industrial targets and so on. The new culture must simultaneously ensure a high level of alignment and the greatest possible autonomy. Although this may seem like an oxymoron, one can find ways of interaction and documentation that allow it. Spotify represents a good example of this (Kniber and Ivarsson 2012).

Redefine teams and their interactions

We have already said that the size of the team matters (it must be small, less than eight people); it should also have all the required skills needed to manage the assigned task. This means that the team should be assigned to a specific mission to be **maintained over time**. The assignment is important for several reasons: (a) it clearly identifies responsibilities for part of the Information Systems from design to maintenance; (b) it enables cross fertilization for roles and skills; (c) it puts all the energy of the Agile in a small context and isolates it from the rest of the Information Systems, thusly reducing the risk of global failures; (d) it enables continuous and asynchronous parallelisms, since the cross dependencies among teams are well known and managed; (e) it creates an easier implementation of the microservices and uses a containerized approach, since a team can be responsible for one or more microservice. To offer a metaphor, let's aggregate the team in a "**Hamlet**". It has its own organization, style, life, economy, rules etc. but, more importantly, it manages a set of microservices. Since the Hamlets have a specific mission, short-term objectives and a medium-long term assignment, we must group them around a larger mission, a specific product, or a set of cross-dependencies. A "**County**" groups one or more Hamlet covering the assigned territory. Within a County, it is easy for individuals to migrate from one Hamlet to another since they operate in similar contexts. Within a Hamlet, individuals that carry out specific activities are identified with a "Role". People with the same Role in one County form a "**Circle**". Each Circle can elect one spokesperson, named "**delegate**"; all the delegates are grouped into a "**Club**" for that Role. We will see the importance of Clubs in the definition of standards and tools. Another spontaneous community of interest is the "**Guild**", where, freely, members can discuss and deepen any subject of their interest; any Hamlet citizen can be part of one or more Guild. A very well defined Guild example is implemented in Spotify (Kniber and Ivarsson 2012).

Redefined concepts behind standards

When you talk about standards, policies and similarities in some of the Agile environments, it seems like you're speaking another language. Standards are sometimes a taboo, but at least, the problem of keeping the teams aligned around common best practices is recognized by everybody; for example, Spotify has defined an approach that they call "cross-pollination" (Kniber and Ivarsson 2012). In large enterprises, especially those in the financial sector, there are a number of regulatory requirements in terms of general IT Governance that you have to fulfil. A smart way of interpreting standards and best practices while keeping the teams motivated and autonomous, is to use the Clubs as the responsible entity for defining standards and best practices for that subject. The exploration of new ways of doing things should be fostered, but when the moment of delivery arrives, if a Hamlet uses different approaches, the subject-matter-expert of the Hamlet and the connected Circle must be ready to explain to the company the reasons behind their decision; if they are good ones, the new approach may be the trigger for a redefinition of the relative standard by the Club. In other words, instead of having an external organizational structure deciding for the developers, we are going to put them at the center of a collaborative decision process. All the interactions within the Circles and the Clubs should be based on collaborative tools to avoid wasting time with never-ending meetings. The Clubs should document the results of the discussions. The risk of the Hamlets' deviation is mitigated by the human nature of conforming to the decisions of the majority (Brown 2000).

Transform documentation in knowledge Management

The documentation is a critical aspect in all the development environments. The creation of detailed and formal documentation can be very boring and smart developers hate boring activities and wasting time. But, on the other hand, a form of knowledge sharing is mandatory to maintain alignment, to avoid repeating stupid mistakes, and to quickly introduce newbies into a matter. For large enterprises, there are also specific regulatory obligations. Each Club should have its own "manifesto" describing in a few pages what they do and how; the document should be associated with videos that should never exceed 5 min in length. All the microservices and interfaces should be documented using specific tools visible to all the Counties. All information should be indexed and searchable by anybody.

The interactions within Hamlets, Circles and Clubs should be based mostly on social development tools.

Architectural key facts embedded in the development process

As we have seen in the dedicated paragraph, there are some technical aspects that are key for keeping the exponential pace; these aspects include the microservices approach, the endorsement of DevOps and Continuous Delivery. They must be part of the way of working from the very beginning. For example, the mission and the artefacts delivered by a Hamlet should be coherent with the microservices concepts. Also, the releases performed by that Hamlet should follow the continuous delivery approach. On the infrastructural side: using containers, we should not be responsible for a container shared between multiple Hamlets. Similarly, the ownership of a microservice should also be assigned to only one Hamlet.

4 Conclusions

The new environment for fostering innovation and enabling the digital transformation is not made by technology only but also by a transforming the enterprise culture and processes, developing solutions that embrace the current mandates of the digital economy, and transforming our information systems accordingly.

From a technology point of view, there is not a single specific element to adopt, as several aspects are strictly interconnected: new capacity for developers, new social development, DevOps, and new architectures (enabling segregation, micro-management, a never-before-seen resilience, standard platforms, shared nothing computing, new data systems, and, last but not least, a loss of association between workloads and our datacenters).

The data are and will always be mostly at the center of operational models and Information Systems. Another example of mutual influence between the shared-nothing architecture is the Big Data and Machine Learning/Deep Learning. They are not new ideas, but in the past, the cost of this type of work was very high due to the computing time and service costs; now the new cost structure of the Big Data and Cloud computing enables this type of research and implementation. Consequently, algorithms are proceeding toward the heart of automatic decision making, probably without completely substituting human workers, but enlarging their capabilities and knowledge.

Organizations will need to abandon the old hierarchical approach to endorse a flexible, semi-autonomous cellular schema, able to foster the agility of small teams, their energy, and their knowledge while, at the same time, ensuring overall alignment with the company's strategy and business objectives. The organizational model should still have the objective of preserving assets and should embed the new technical architectural principles directly into its rules (e.g., 1 team manages 1 microservice).

In the end, a new environment is created with new ways of thinking, new ways of doing, new ways of spending, and new ways of dealing with customers.

This is not a fantasy-adventure: companies like Google, Facebook, Amazon, Twitter, Spotify, Netflix and many others have had these approaches in place for many years, setting the standard for what customer experience and IT efficiency should be.

The others will have to catch up quickly!

References

A Rajendra (2014) Big data computing. CRC Press 2014, Taylor & Francis Group, Boca Raton. ISBN 978-1-4665-7838-8

Blanter A, Holman M (2016) Internet of things 2020: a glimpse into the future. Available at Kearney https://www.atkearney.com/documents/4634214/6398631/A.T.+Kearney_Internet+of+Things+2020+Presentation_Online.pdf/af7e6a55-cde2-4490-8066-a95664efd35a Accessed 28 Sept 2016

Barroso LA, Clidaras J, Hölzle U (2013) The datacenter as a computer an introduction to the design of warehouse-scale machines, 2nd edn. Morgan & Claypool www.morganclaypool.com ISBN: 9781627050104

Brown R (2000) Psicologia social dei Gruppi. il Mulino, Bologna, titolo originale Group Processes. Dynamics within and between Groups. Oxford, Blackwell Publishers, ISBN 88-15-10669-3

Cisco (2014) Linux Containers: Why They're in Your Future and What Has to Happen First. Available at http://www.cisco.com/c/dam/en/us/solutions/collateral/data-center-virtualization/openstack-at-cisco/linux-containers-white-paper-cisco-red-hat.pdf Accessed 8th Sept 2016

Cohen D, Sargeant M, Somers K (2014) 3-D printing takes shape. Available at http://www.mckinsey.com/business-functions/operations/our-insights/3-d-printing-takes-shape Accessed 30th Sept 2016

Columbus L (2015) Roundup of 3D Printing market forecasts and estimates. Available at http://www.forbes.com/sites/louiscolumbus/2015/03/31/2015-roundup-of-3d-printing-market-forecasts-and-estimates/#128b076e1dc6 Accessed 28th Sept 2016

Container (2015) The shipping container a complete history available at http://www.containerhomeplans.org/2015/03/a-complete-history-of-the-shipping-container Accessed 8 Sept 2016

Docker (2016a) Docker team. Available at https://goto.docker.com/rs/929-FJL-178/images/Docker-Survey-2016.pdf Accessed 8 Sept 2016

Docker (2016b) Docker team. Available at https://www.docker.com/sites/default/files/WP_ModernAppArchitecture_04.12.2016_1.pdf Accessed 8 Sept 2016

Earls A, Baya V (2014) The road ahead for 3-D printers Available at http://www.pwc.com/us/en/technology-forecast/2014/3d-printing/features/assets/pwc-3d-printing-prototyping-finished-products.pdf Accessed 29 Sept 2016

Feinleib David (2014) Big data bootcamp what managers need to know to profit from the big data revolution. California, ISBN: 978-1-4842-0040-7

Ford KM (2001) Cognitive Prostheses. Institute for Human & Machine Cognition University of West Florida. Available at http://www.lpi.usra.edu/publications/reports/CB-1089/ford.pdf Accessed 20 Sept 2016

Fowler M (2005) The new methodology. Available at http://www.martinfowler.com/articles/newMethodology.html Accessed on 25 August 2016

Fowler M, Lewis J (2014) Microservices a definition of this new architectural term. Available at http://martinfowler.com/articles/microservices.html Accessed 8 Sept 2016

Hinssen P (2010) The new Normal Explore the limits of the digital world. Mac Media NV, Belgium. ISBN 978-9-081-32425-0

IDC (2014) EMC Digital Universe with Research and Analysis by IDC. The digital universe of opportunities: rich data and the increasing value of the internet of things. Available at http://www.emc.com/leadership/digital-universe/2014iview/executive-summary.htm Accessed 15 Sept 2016

Ismail S (2015) Exponential organizations. Marsilio Nodi Venezia, ISBN 978-88-317-2179-0

Kang J-M, Lin T, Bannazade H, Leon-Garcia A (2014) Software-defined infrastructure and the SAVI testbed. Network and Mobility Lab, HP Labs., Palo Alto, Available at http://citeseerx.ist.psu.edu/viewdoc/download?doi=10.1.1.686.3439&rep=rep1&type=pdf Accessed 2 Sept 2016

Allan Kelly (2008) Changing software development, learning to become agile. John Wiley & Sons, Chichester. ISBN 978-0-470-51504-4

Kniber H, Ivarsson A (2012) Scaling agile @ spotify with tribes, squads, chapters & guilds. Available at https://ucvox.files.wordpress.com/2012/11/113617905-scaling-agile-spotify-11.pdf Accessed on 24 August 2016

Kniber H (2014) Spotify engineering culture. Available at https://spotifylabscom.files.wordpress.com/2014/03/spotify-engineering-culture-part1.jpeg Accessed on 24 August 2016

Koff William and Gustafson Paul (2012) 3D printing and the future of manufacturing. Available at http://assets1.csc.com/innovation/downloads/LEF_20123DPrinting.pdf Accessed 28 Sept 2016

Krishnan Krish (2013) Data warehousing in the age of big data. Morgan Kaufmann, Waltham. ISBN 0124058914

Levinson M (2006) The box: how the shipping container made the world smaller and the world economy bigger". Princeton University Press ISBN: 9780691136400

Bertrand Meyer (2014) Agile! the good, the hype and the ugly. Springer, Zurich. ISBN 978-3-319-05154-3

Soumendra Mohanty, Jagadeesh Madhu, Srivatsa Harsha (2013) Big data analytics. Springer, New York. ISBN 978-1-4302-4872-9

Fernando Monteiro (2014) Learning single-page web application development. Packt Publishing Ltd., Birmingham. ISBN 978-1-78355-209-2

Raj P, Jeeva S, Singh Chelladhurai Vinod (2015) Learning docker. Packt Publishing Ltd, Birmingham. ISBN 978-1-78439-793-7

Pollyanna Pixton, Paul Gibson, Niel Nikolaisen (2014) The agile culture leading through trust and owernership. Addison-Wesley, London. ISBN 978-0-321-94014-8

Price Daniel (2015) Available at http://cloudtweaks.com/2015/03/surprising-facts-and-stats-about-the-big-data-industry/ Accessed 2 Sept 2016

Qrimp (2008) The difference between IaaS and PaaS. Available at http://www.qrimp.com/blog/blog.The-Difference-between-IaaS-and-PaaS.html Accessed 7 Sept 2016

Reactive (2014) The Reactive Manifesto web manifesto with 16000 signers, http://www.reactivemanifesto.org Accessed 30 Aug 2016

Schwaber K, Sutherland J (2012) Software in 30 days: how agile managers beat the odds, delight their customers, and leave competitors in the dust. Wiley

Schmidt E (2010) Every 2 days we create as much information as we did up to 2003. Available at https://techcrunch.com/2010/08/04/schmidt-data/. Accessed 1 Sept 2016

Stonebraker Michael (1986) The case for shared nothing. University of California Berkeley, Ca., Available at http://citeseerx.ist.psu.edu/viewdoc/summary?doi=10.1.1.58.5370 Accessed 2 Sept 2016

Conceiving and Implementing the Digital Organization

Mariano Corso, Gianluca Giovannetti, Luciano Guglielmi and Giovanni Vaia

Abstract The digital transformation is affecting numerous industries, including media, retail, automotive, transportation, and healthcare. To address the modern challenges of digital technologies, increasing customer demands, and a rapidly changing enterprise setting, CIOs and their IT organizations are required to extend their performance profile, adopting a new organizational design and a broader range of behaviors. Today, CIOs are learning that they cannot manage current strategic, operational and investment responsibilities within existing company boundaries while still ensuring stability and hard cost management. IT organizational units should move beyond their current focus on operations and systems and adopt new behaviors, to facilitate and lead new digital innovations, seize new opportunities, and raise business performance in the marketplace. CIOs recognize that they must remove historical and legacy commitments to create new connections in processes, structures and roles and to reinvigorate IT value potential in a new digital environment. They need to redesign internal organizational interactions to liberate resources, time and attention dedicated to new challenges. They also need to move beyond enterprise boundaries and navigate new ecosystems to search for innovation opportunities. This chapter presents insights on how CIOs must orchestrate

---

The views expressed in the paper are those of the author and do not necessarily reflect those of the company.

M. Corso
Politecnico di Milano, Milan, Italy
e-mail: mariano.corso@polimi.it

G. Giovannetti
Amadori, Cesena, Italy
e-mail: gianluca.giovannetti@amadori.it

L. Guglielmi
Gruppo Mondadori, Milan, Italy
e-mail: luciano.guglielmi@mondadori.it

G. Vaia (✉)
Ca' Foscari University of Venice, Venice, Italy
e-mail: g.vaia@unive.it

© Springer International Publishing AG 2018
G. Bongiorno et al. (eds.), *CIOs and the Digital Transformation*,
DOI 10.1007/978-3-319-31026-8_10

innovation across functions and external networks to reinvent and structure value delivery and new business models, stimulating a wider and more productive conversation with different actors. This chapter also argues that CIOs need to combine and harmonize physical and digital, thusly breaking from traditional processes and establishing new digital methods of working.

1 Introduction

Our fast-paced world is in a state of constant change. New digital technologies are making our lives easier, by shaping and changing the way in which we interact, work and purchase. What was new yesterday is old today and keeping up with new trends and transformations in the market is harder than ever before.

Consequently, new resolute digital entrants are disrupting established traditional business models and making a long term competitive advantage impossible to attain. New digital players emerge in seemingly stable industries: from health care to finance, from transportation to manufacturing. Think about Uber, Facebook or Alibaba. Each of these companies didn't simply change its respective industry but rather created a completely new one. They were able to translate a vision into a reality and were bold enough to take the risk and change the world. In the past, nobody would have ever expected them to succeed, but they actually did.

Established companies have a choice to make: disrupt themselves or let someone else do it. Think about Netflix for example. Today, the world's largest online movie and TV show streaming provider, with more than 83 million subscribers worldwide, started as a DVD rental service via mail in 1997. The company was a new player in the market dominated by Blockbuster—a giant video rental company that seemed to be unbeatable at the time. Over time, Netflix started to realize that something was changing in the market: new technologies were emerging and were changing the consumers' expectations and needs. Netflix recognized a huge opportunity in this change so they decided to reinvent themselves, innovate their business model and eventually transform into the media streaming company we know today. What about the giant Blockbuster? It doesn't exist anymore. They were not able to recognize the faint signals of change in the market so when the digital revolution arrived it was too late for them. This is basically what happened to many companies all around the world. As a result, half of the companies on the Fortune 500 list in 2000 have now fallen off because they failed to adapt to the digital age (Wang 2015).

Therefore, in a world in which everything changes so fast, an advantage can be gained only through constant innovation and a deep study of market trends and dynamics. To borrow the term introduced by PwC (2016) to describe this phenomenon—the world is "in beta". In this "beta" world, companies are required to re-organize themselves and their systems in order to become more flexible and able to regularly change the way in which they create value for their consumers.

2 The New Digital Enterprise

A digital enterprise can be defined as an enterprise that has the following four characteristics:

(1) a company is able to easily adapt to the "beta" era;
(2) a company sets ambitious goals and constantly challenges the status quo;
(3) a company's decision-making is driven by innovative thinking and relies on a deep knowledge of the external market;
(4) a company uses technology in order to maximize both the strategic and the operational value and gain a significant competitive advantage.

First, a new digital enterprise needs to deeply understand the period of time in which it operates. The world is fast-paced and new consumers, together with new technologies and opportunities are arising every single day. As a result, the level of complexity increases and companies must make decisions that will affect their future in an environment in which uncertainty, volatility and ambiguity are the predominant characteristics. In this new business landscape, companies have to keep on "experimenting and learning, [to] identify new opportunities, exploit them fast and move on".[1]

Second, new digital enterprises are ambitious as they firmly believe that their idea can change the way in which other people in the world live or behave. They see opportunities that arise when they challenge the status quo. This is exactly what Amazon is doing, for example. Think about its ideas for a new delivery network: from Amazon logistics services, to the recent intention to open a Parcel Locker Service across all of Europe, along with its commitment to increase the "On-demand" delivery services. Amazon keeps on challenging the status quo and tries out new solutions in order to maintain its advantage in the market.

Third, new digital enterprises are making their decisions quickly based on their interpretation of the ongoing changes in the business environment. In order to compete in these new speed-based economies, the organization must be able to adapt to this speed and quickly make the right decisions. The decisions are therefore driven by innovative thinking combined with a deep knowledge of the external environment of reference and its elements. The main elements are: consumers, new technologies and competitors. In the new digital enterprise, the consumer is at the center. Changes in his/her needs and/or behavior are constantly monitored and he/she is the main driver of all the decisions the company makes. New technologies enable companies to recognize the weak signals of change in the market and, therefore, in the consumers' needs and behaviors. The new digital organization can and must understand its consumer very well and anticipate his/her needs. Doing so is imperative for organizations that seek to succeed in this new market. As Steve Jobs once said: "Get closer than ever to your customers. So close that you tell them what they need well before they realize it themselves." This is not possible without a deep

---

[1]http://www.worldinbeta.com/introduction.

knowledge of the available technologies. It is hard to create great innovative products with old technologies. Organizations have to keep studying the new technologies in the market and combine them with the established ones; this practice can give different perspectives and make the work easier and faster.

Another important element that needs to be constantly considered in the market is the competition. Benchmark is everything and can be very useful in gaining different perspectives and new ideas. In today's world, benchmarking is no longer referred to as the study of the direct competitors; instead, organizations have to broaden their horizons and study what other companies in different sectors and markets are doing. Indeed, all other industries and organizations can be a source of inspiration. Consider the example of ING Netherlands. ING is a Dutch multinational banking and financial service corporation. In 2014, they decided to change their business model and become an omni-channel bank. To do so, they decided to follow the example set by Spotify, a Swedish music, podcast and video streaming service that provides digital rights management-protected contents and that reached 30 million paying subscribers in March 2016 (data retrieved from www.statista.com, october 2016). ING spent a considerable amount of time studying the way in which people work inside Spotify; they understood that the secret was a more connected organization in which people from different departments work together. So ING, in transforming its organization, learned from a company in a seemingly unrelated industry and re-organized itself in order to adopt this new way of agile working.

Finally, the new digital organization understands the potential of new technologies in creating both operational and strategic value. The new technologies on which it bases its innovation process are mainly mobility, social networks, cloud computing and data analytics. These four forces can provide an important advantage to the company. Even though each of them is powerful by itself, their collective use can provide strong synergies (Rameshkumar 2013). Gartner defined these synergies as the Nexus of Forces. These, combined with new emerging technologies such as the Internet of things, 3D printing, wearable and smart machines can create endless possibilities for innovation. The technologies that are recognized as valuable for the organization are combined with the already established ones and are a main source of innovation. In this new type of organization, it is clear that the technologies are the key to providing additional strategic value to the company's decisions and actions. Therefore, in this organization the IT department is at the center and collaborates with all the other departments in order to increase the created value and reach a market advantage.

What is important to realize is that already established companies may feel threatened by these new digital entrants, and are learning from them how to change and where the market is going. They are therefore trying to re-define themselves, their strategy and the way in which they work and reach the consumer. A very interesting case in this sense is Walmart. Walmart is an American multinational retail corporation that operates in discount department and grocery stores. It is the world's largest company by revenue in 2016, according to the Fortune Global 500 list. At the beginning of this year, Walmart announced that it will close 269 stores in the US and, just a few weeks ago, in August 2016, they announced a deal to acquire the e-commerce website Jet.com for US $3.3 billion. They clearly recognized the threat that Amazon is becoming in the market, and that the grocery shopping industry will have a completely different look in a few years. They also recognized that they were not strong enough in e-commerce, so they decided to both re-define their strategy and redesign their company.

3 Adapting the Context to the Digital Age

Some believe that innovation comes easier in a start up than in an established business. The reasoning behind this notion is that, the bigger the organization, the harder it is to transform it. It is like a big boat in the harbor. To go out or change directions it takes time, while small ships can easily move and change their route. Established corporations, however, have some other advantages such as the assets, resources and capabilities necessary to fuel innovations (Govindarajan 2016). Indeed, many big organizations showed they were able to successfully innovate and succeed in the business transformation journey. Consider, for example, Cisco, the largest networking company in the world. In 2013, they were the market leader but they saw that something around them was changing; new technology trends were arising such as the cloud, the Internet of Things and mobility. They decided it was better to take the risk and innovate their company before it was too late. They started a transformation journey and disrupted their business model before others could do it. Liz Centoni, Senior Vice President and General Manager of Computing Systems Product Group at Cisco affirmed that the key to success in this transformation was to understand that the market was changing and change with it while they were strong. They re-designed their business model with the intention to simplify their products and the customers' lives.

Cisco worked to re-design the organization; they created multi-disciplinary teams and defined a clear set of actions to transform the company. Another example of this is Starbucks, the American coffee company. They anticipated and began to invest in mobile technology many years before others. Today, they are considered a prime example of a customer- centered mobile and retail experience for the forthcoming years.

Thus, established companies can succeed in the transformation. But what should be understood is that transforming a company's business model demands a change deeper than merely providing the latest digital technologies. A company transformation is an evolutionary path that involves people, new business models and technologies. Together, these three elements give new perspectives and a new vision, which are at the same time both radical and concrete, to the organization. The organization business transformation can be divided in four main steps (Fig. 1):

1. Studying the external environment;
2. Getting the right Vision;
3. Designing the Journey;
4. Executing.

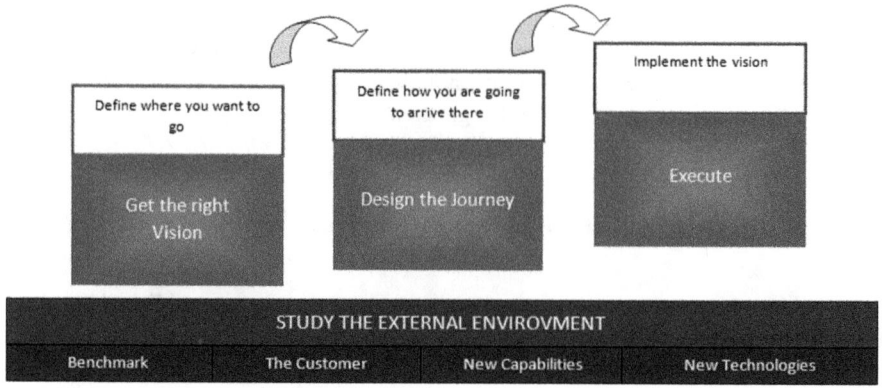

Fig. 1 The business transformation

The first step when facing a business transformation is to *study what is going on in the external market*. This is the only step that the company has to do periodically. Indeed, this can be very helpful for understanding when a new transformation occurs. What needs to be studied in the external market can be summarized in four main elements: the competitors, the consumers, the labor market and the new technologies. Studying what the *competitors* are doing today is deeply different than what it was years ago. The benchmark has to be done with different companies, from different sectors that sell different products. All the companies that are succeeding in the market have a lesson to offer, as illustrated by ING Netherlands taking Spotify as an example for its business transformation.

The second very important element that the company has to monitor constantly is the *customer*. In this new era, the customer is at the center of everything within a company. Creating his/her journey is imperative to understanding the way in which he/she interacts with the company. According to Gartner (2014), creating the best customer experience is the key factor for helping the company grow. The organization should know everything about its consumer: his/her consumption habits, characteristics, thoughts and feelings about specific products. Thanks to the help of new technologies, this is easier than in the past. If the organization doesn't understand its consumer in this new era, all its other efforts will be vain.

The third element that the company has to monitor during the business transformation is the *labor market*. Since the company is transforming itself and the way in which it operates, new capabilities are required. Understanding which of these are offered in the labor market is important when hiring and developing the best talents; companies also need to always remember that it is hard to do memorable things without valuable people.

Finally, the organization has to figure out how *new technologies* in the market can be used as new opportunities to increase the organization's value creation. The new technologies can create new competitive advantage for the organization.

Being conscious of the new market trends and solutions is essential for both the operations and the strategy.

The second step is to *get the right vision*. Deciding how the company will look in the future is crucial for the success of its business. Organizations that use technology as a strategic asset are succeeding in this era, but it is important to be conscious of the fact that strategy, not technology, drives the business transformation (Kane et al. 2015). Some may argue that the right decisions are hard to make in a time in which the market is constantly dealing with uncertainty, volatility, ambiguity and complexity. Defining new ways through which the organization can create additional value and build a sustainable market advantage is essential to building up the foundation for creating a new, solid organization. Where does the transformation journey take the organization? How does the organization want to transform itself? Where will the organization play in the future? How will the organization gain competitive advantage? Answering these questions is essential for creating an effective new organization vision and a new transformation journey. Doing it right is difficult but not impossible, and this is why the vision should be decided only after a deep study of the external environment. Starting with the data, the trends and consideration of the practices of other companies, together with your consumers' preferences is very important for defining where you want to go in the future and surviving the digital age. Of course, organizations have to accept that they can't know everything, but they can experiment with new possible solutions by combining the company's traditions and a new, evolved culture.

The third step is to *design the transformation journey*. Once the company defines where it wants to go, it is time to define how to arrive there. In this step, the organization has to define its new business model, the transformation leader, the leadership team and the new capabilities that it needs to develop. It is a delicate moment for the company because it starts planning and defining its new path.

In order for the transformation to succeed, there are five main steps that an organization needs to define and consider when building up its new transformation journey. These steps are (Fig. 2):

a. Defining the transformation leader;
b. Using the IT department as a transformation enabler;
c. Considering the HR department as a transformation facilitator;
d. Revolutionizing the way of working: adopting multi-functional teams;
e. Coordinating the transformation from the center: Establishing the EPMO.

The fourth and last step is the *execution*, the time in which implementation occurs. It is considered the hardest step because it is the moment in which the real change happens in the company. Most companies are actually able to set the right vision but then fail to execute it.

Fig. 2 The Transformational journey

4 Organisational Changes to Succeed the Transformation

4.1 The Transformation Leader

The most delicate step in this process is identifying the transformation leader, since he/she will be the one contributing most to the transformation's success. While one might expect the Chief Executive Officer to lead the transformation, it is essential to note this is not always the case. According to an Accenture report (2015) the transformation ownership is actually divided between the CEO (38%), CIO (33%), CDO (17%) and the CMO (8%). The reason that the percentage of the CEOs and CIOs that lead the transformation is so similar is because the IT department is becoming more and more relevant in the new digital organization. In fact, today, IT has become the core of the digital enterprise value since it can bring different functions to a new technological level. Take for example, the case of Amadori, one of the leading companies in the Italian agro-food sector. In 2016 they launched their new transformation plan, seeking to change and remodel the way in which the organization creates both value and efficiency. In their case, the transformation plan is under the responsibility of the Chief Information, Process and Business Transformation Officer since the IT function has an inter-function operational and strategic element.

Indeed, understanding the impact that the transformation plan has on the organizational strategy is important but what is even more important is an understanding of its impact on the operations. If what you have defined in the strategy is not reachable by the operations, it means that everything you have done is in vain. After the transformation leader has been chosen, the organization has to build a transformation team that will help the leader organize, orchestrate and drive the business transformation. We are therefore talking about large companies in which it is very hard for a single person to have full control. Hence, creating an extended group can be very useful for always being in control of the different impacts that the transformation has on the organization.

4.2 The Information Technology Department—From the Periphery to the Center

4.2.1 The Evolution

With the rise of the business transformation era and the new digital enterprises, IT has become the central function for many organizations. As we saw in the last year, more and more companies shifted from traditional organization to new digital enterprises. They started to realize the opportunities that IT governance had to offer to business in terms of value and differentiation. Therefore, investing in IT has become a fundamental business imperative for companies (Peterson 2004) in order to face the transformation and be able to survive in the market. The IT function has changed from a position of passive "order taker" to an active one that leads the business transformation and has power in business decisions for both operations and strategy. Interestingly, the name of the function itself is changing as well: what was once called an "EDP Centre" was later replaced by the "IT Department", and then by the "Information Systems Function", followed by the "ICT business unit", and today, it is called the "Digital Innovation Business Unit".

Few functions have steadily evolved like ICT over the past twenty years, not only in terms of designation, but also in terms of organizational models, roles and missions. The ongoing debate as to what lies ahead for the business unit and its managers provides us an outlook of a function having an identity crisis, constantly poised precariously between opposing tensions.

In fact, on one hand, we are witnessing a growing effort to industrialize technology and *make ICT a commodity*, both usable as a service and also manageable as a utility that the business has no need to worry about, given the existence of robust and ongoing support with ever better performance at decreasing costs. The technological evolution of hardware and devices linked to the so-called Moore's law,[2] and the development of an increasingly standardized, global and competitive market offer seem to naturally accompany this tension. The interpretation of this trend has generated a line of thought which has found its manifesto in Nicholas Carr and the so-called Utility Computing. In 2003, with the controversial article "IT

---

[2]In 1965, Gordon Moore, future co-founder of Intel, based on empirical observations, suggested that the number of transistors in microprocessors would double in approximately every 12 months. Although the times have been gradually extended to 18 or 24 months this prediction, which came to be known as the first law of Moore, proved to be substantially correct for over three decades, conditioning the evolutionary roadmap of microprocessor manufacturers and consequently, of hardware and device manufacturers.

Doesn't Matter" published in the *Harvard Business Review*, Carr argued that, precisely because of its progressive dissemination and "standardisability," the strategic importance of information technology in the world of business was destined to decrease, with obvious consequences in terms of the necessary progressive outsourcing of IT activities towards specialist suppliers. From the internal organization perspective, this means placing emphasis on efficiency, centralization and standardization, and leveraging the development of an increasingly "commoditized" offer to proceed to a progressive streamlining and the outsourcing of activities.

On the other hand, however, a second, apparently opposite trend has gradually emerged: ICT is also a source of innovation for business. An often radical innovation which, even in sectors once distanced from technology, represents a key foundation for the transformation of products, processes and business models. In order to remain competitive, it is therefore essential to manage the ICT function like a research laboratory that is open to the market, and deeply connected with the business. From an organizational perspective, the emphasis should be placed on factors such as innovation, proximity to the core business and stimulus for creativity. Sourcing practices should therefore not be carried out as a type of simple delegation or the outsourcing of non-core activities, but as opportunities to access innovative competences that are carefully selected on the basis of parameters that go far beyond the price, and are integrated with internal resources.

Gartner (2014) provides a more detailed historical overview of how the role of IT has changed over the decades. According to this model there are three eras (Gartner 2014):

- IT Craftsmanship era;
- IT Industrialization era;
- IT Digitalization era.

In the *IT Craftsmanship era*, the IT department was isolated from the other departments and was considered a passive order taker. Indeed, people working in IT were mainly taking orders from other functions, doing exactly what other people wanted them to do and rarely pushing back (Westerman et al. 2016). The focus was on programming and system management.

As time passed, the potential of IT was recognized for what it was. Organizations started to make more of an effort to link IT to other functions, and switch from a one-way to a two-way communication. This was defined by Gartner as the second era of IT: *the IT industrialization era*. The value of IT started to be recognized and connected to its ability to improve business performance. Therefore, IT started collaborating with functions and colleagues from other departments, initially treated as customers. The idea was to link technology to processes in order to increase their efficiency and effectiveness. By studying the processes and linking them to new technologies, IT became ever more conscious about what was going on in the enterprise, allowing IT staff to have a deeper knowledge of the overall organizational processes and structure.

Conceiving and Implementing the Digital Organization

IT is now taking a step ahead. We are now in the era that Gartner defines as *the digitalization era*, in which IT is characterized by interdisciplinary approaches that lead to greater collaboration between different figures in the company. In the previous era, the colleagues were treated as clients, while now they are IT business partners that collectively shape the company's future. IT is increasingly becoming the center of organizations that want to be more competitive in the market. According to SAP (2016), in this new era, IT is a core element in the organizations' value chain. Indeed, IT no longer focuses only on technology, but also on processes as well as on business models.

4.2.2 CIO as Chief Innovation Officer

As the role of the IT department evolves within organizations, the responsibilities of its leader grow. In addition to managing the IT department, the CIO is asked to orchestrate a new, different group of business elements. Indeed, in many cases he/she is the one in charge of the organization's transformation and is charged with coming up with new solutions for the business. Of course, the role that the CIO and his/her department has in the transformation journey depends a lot on his relationship with the organization CEO as well as with the other board members. Therefore, if the CEO doesn't recognize the power that the IT department and its leader have in the organizational transformation, then the CIO freedom of action will be drastically lowered. On the other hand, this is also a responsibility for the CIO, who has to be able to conquer the new position within the organization. Indeed, he/she has to be the one who works to change the idea that people within the organization, especially the board, have about an IT department, to insure they recognize its potential. The shift that the CIO is required to do is important. He/she must no more be considered just the technical person but the board member that can give a very important contribution to reach a significant market advantage. He/she is now asked to be an active business partner for the different functions. Indeed, thanks to his/her 360° view of the organization, he/she is able to give a significant contribution to shape the organizational strategic decisions and the future of the business.

In the end, the business transformation is an incredible opportunity for the CIOs. As for the organizations, the CIOs who are unable to adapt to this new way of working and challenge themselves to transform their role, will be brutally disrupted by the ones who could. Therefore, according to Gartner (2016), there is no future for the old type of CIO that was mainly IT oriented and largely self-referential. In the next years, there will be a much more dramatic distinction between CIOs who adapt to the transformation and those who don't. Those in the latter category would be well-advised to adapt their habits, as it may soon be too late.

4.2.3 Ensuring Day by Day Operations and Focusing on Growth

As we saw, the IT department is reshaping its way of working and re-defining itself within the new digital enterprise. Gartner defines a combination of two different roles within the IT Department as "bimodal IT", since two significantly different ways of working coexist and collaborate within a single IT department.

Bimodal IT[3] approach recognizes the need to bring together two modes, referred to as *mode 1* and *mode 2*, coherent and separate for managing IT, with one focused on stability and the other on agility. Mode 1 requires rigor and respect for sequentially emphasizing security and precision, while mode 2 instead requires an exploratory and non-linear approach that emphasizes agility and speed. Both modalities of working are indispensable to an organization's value creation and success in the digital transformation.

Mode 1 is the part of the IT department that takes care of traditional, routine IT operations. It is by definition more plan-oriented and focuses on daily activities. For this reason, it is normally IT-centric and works to ensure support to the daily users and the application evolution. Mode 1 is very plan driven; as such, its cycle time is usually long and its projects usually take several months to be completed.

Mode 2 is the part of the IT department that deals with business process transformation, new digital skills and delivering new strategic projects and solutions. Consequently, it is more business-oriented and it focuses on organizational growth. It typically delivers solutions in a very short period of time, such as weeks or even days. This mode is therefore more exploratory, agile and risky.

The need to accommodate both modalities of work within the same IT department has spurred the development of a number of contingency models. Such models are developed on the basis of actual companies' experiences and are aimed at proposing more suitable organizational forms depending on the prevalence of certain requirements. The most famous of these contingency models is Nolan and McFarlan's (2005): by plotting the need for robustness (the need for reliable IT) on the vertical axis and contribution to innovation (the need for new applications) on a horizontal one, the authors identify four different modes of IT governance (Factory, Support, Strategic and Turnaround). The choice of a particular mode depends on how much a company relies on IT for maintaining its cost-effective back-end systems as opposed to aggressively competing on the market with cutting-edge technology solutions.[4]

More recently, the Politecnico di Milano School of Management (Osservatori 2016) has proposed a contingency map which, by juxtaposing the Role of ICT

---

[3]Aron D, McDonald M (2013) Taming the Digital Dragon: The 2014 CIO Agenda. Gartner Group. See also: Gartner 2015, IT Glossary—Bimodal IT. http://www.gartner.com/it-glossary/bimodal.

[4]McFarlan, F. W. and Nolan, R. (2005), Information Technology and the board of directors, *Harvard Business Review*, Vol. 83, No. 10, pp. 96-106.

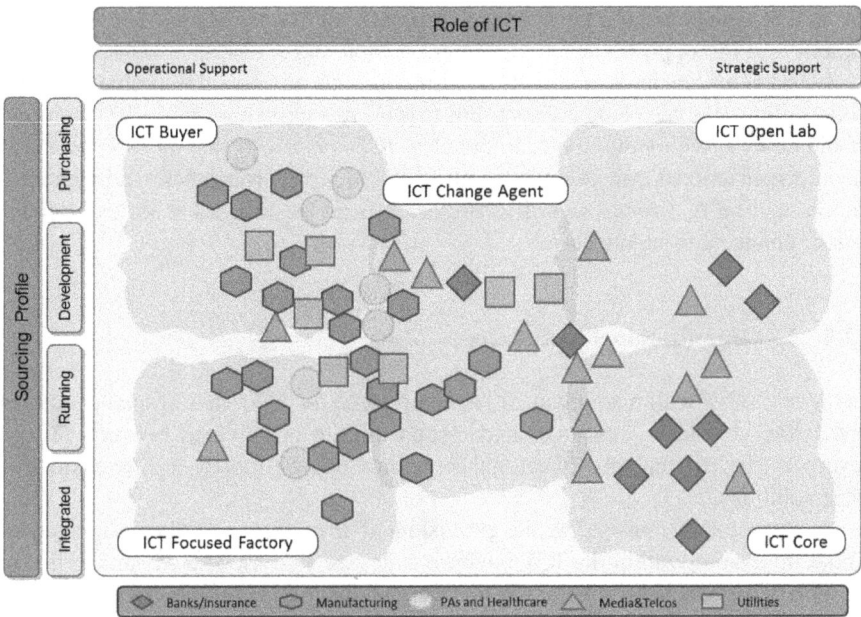

Fig. 3 Polimi ICT models map

(Operational or Strategic) with the Sourcing Profile (integrated, management,[5] development[6] or purchasing), enables the mapping of five clusters (or ICT models) with homogeneous characteristics in terms of organisational structure, governance model, sourcing strategies and the internal and external key competencies required (Fig. 3).

All the contingency models share the limit to indicate a prevailing solution that is empirically "satisfactory" and represents the best possible compromise in light of the company's situation. In medium-high complexity situations, however, different elements and needs co-exist within the same organization, making it difficult to define a prevalence. Moreover, due to the changing evolution of competitive conditions, business needs are changing and are rapidly making the current ICT model inadequate, forcing it to "swing" between different organizational models. In this regard, one can speak of a veritable "pendulum" in which ICT models oscillate

---

[5] An organisation follows a "management" sourcing profile when it is oriented towards retaining management and maintenance operational activities internally and by delegating to suppliers activities related to development, for which it considers itself to be unable to ensure the correct level of professional and competitive updating.

[6] An organisation follows a "development" sourcing profile when it is oriented towards retaining development activities internally and by delegating to suppliers those tasks related to management and maintenance for which it considers itself unable to ensure the correct level of efficiency and competitiveness.

due to the changing balance between the needs of "consolidation and standardization" and the opposite needs of "differentiation and speed of response".

Hybrid solutions have emerged over time in order to solve this problem which, while failing to find a single prevailing model, recognize that more ICT can and should co-exist within the same company: an *ICT commodity* that must be managed by an organizational unit capable of ensuring efficiency, robustness and standardization, and an *ICT Innovation* that, protected from the first, must strive for innovation, change and uniqueness.

4.2.4 IT Organizational Structures

As the role of IT within an organization changes, so do the organizational structures that define how tasks, resources and responsibilities are divided between IT and business. The existing organizational structures can be mostly attributed to five basic solutions.

A first solution consists in the provision of a unit specifically designated to digital innovation within the ICT business unit (see Fig. 4). The advantage lies in recognizing a specific competence for innovation within the ICT Department to which ad hoc resources can be dedicated in an integrated approach towards ICT management and under the supervision of its manager. The risk of this solution, advocated and adopted by many CIOs, is that, in retrospect, the commitment and the resources dedicated to the development of innovation ends up being marginal in terms of the commitment in the operational management of ICT. The same CIO in charge will end up in a tense situation in which the desire to be seen internally as the strategic business contact for digital innovation is frustrated by the fact of being, above all, viewed as the bearer of constraints for the continuity, security and robustness of the current operational processes. The solution may function adequately in the presence of an operational machine for current management that is sufficiently streamlined, and better if supported by an effective outsourcing policy.

A second solution, a dual solution, assumes that the ICT business unit is a sub-unit of a first-level department designated to Digital Innovation (see Fig. 5). The advantage, with respect to the first solution, lies in giving greater organisational weight to innovation, which becomes the main mission of the Chief Digital Officer responsible at the first level. The difficulty lies in the delicate balance between the CDO and the manager of the ICT business unit; the latter ends up in a subordinate role while remaining in direct control, in most cases, of most of the department's

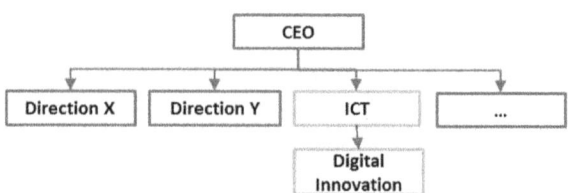

Fig. 4 Digital Innovation as a unit inside the ICT department

Fig. 5 ICT as a unit inside the Digital Innovation department

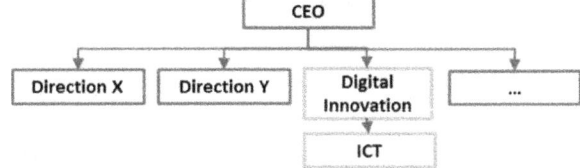

Fig. 6 ICT and Digital Innovation as separate independent units

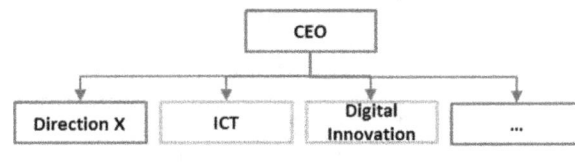

Fig. 7 Digital Innovation as a unit within a business unit separate from ICT

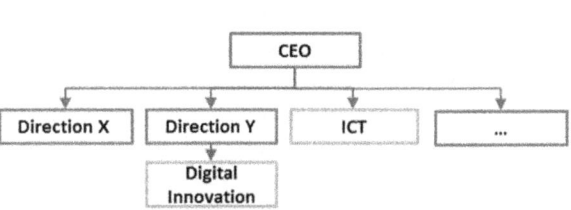

resources. This solution works in the presence of the strong credibility and authority of the Chief Digital Officer within the business, and "non-intrusive" support to the manager of the ICT business unit.

A third solution which attempts to overcome the disadvantages of the first two is providing two directions at the organization's first level: ICT and Digital Innovation (see Fig. 6). The advantage of this solution is in protecting innovation, especially in the medium-long term, without suppressing it with urgent issues of an operational nature in terms of resources, time and managerial attention. The risk, by contrast, lies in the potential distancing of the Digital Innovation unit's staff from the specific responsibility of managing the implementation and operation of digital solutions. This threatens to lead to a progressive detachment from reality, while the loss of the role with respect to innovation is likely to frustrate and deplete the professionalism of the ICT Department's staff. The key issue in this case lies in designing ways in which these two departments transversally interact in the inception and development of digital innovation projects, and would be better if suitably reinforced by job rotation policies for people between the two departments.

A fourth solution consists in providing a Digital business unit reporting to a parallel business unit, independent from the IT Department (e.g. Marketing or Supply Chain); this unit would be tasked with monitoring the latter's operations (see Fig. 7). The advantage lies in the convergence between the Digital Innovation unit and the business, or at least the part of the business in which the Digital unit is managed. The disadvantage lies in strongly polarising digital innovation on a specific business component thus distancing it from the ICT business unit and from

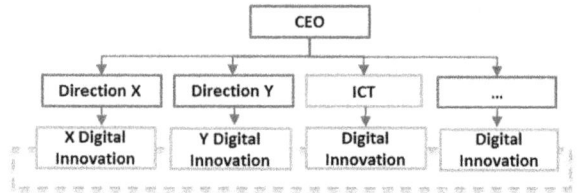

Fig. 8 Digital Innovation as a network of interrelated units inside different business units

the rest of the organisation. This solution may work when the characteristics of the C-Level manager of the business unit in which the Digital Innovation department is placed are of a high authority and system vision and when there is also a clear focus on digital innovation priorities.

A fifth and final solution is providing roles or small teams designated to digital innovation in the main line business units, thus providing functional relationships and transversal project teams in collaboration with a unit geared towards innovation within the ICT business unit (see Fig. 8). The advantage is the contiguity of digital innovation. The limitation is the complexity and the difficulty in maintaining an overall consistent strategy and architecture. The operating conditions include a strong sensitivity to digital innovation by top management and all the C-levels and a notable pervasiveness of digital competences within the business.

None of the five solutions outlined above are free from limitations and, in complex organizations, several of these solutions co-exist, often without a clear predominance of one over the others.

Therefore, in the face of increasing pervasiveness and the strategic importance of digital transformation, organizations are searching for IT models that can combine security, efficiency and operational robustness that are typical of enterprise-class IT, with the flexibility and the innovation capacity of a Start-up and the pervasiveness, usability and customer proximity of a consumer service. A seemingly impossible challenge which cannot be addressed with a traditional organizational solution that views IT as Organizational Silos—or a set of organizational silos—that are distinct from the business. As Bernard Golden, the Thought Leader of Cloud computing wrote, IT is changing from "support the business" to "be the business". The new IT models are, therefore, not to be interpreted as independent organizational units, but as focal elements of a digital innovation widespread network that involves the entire organization and extends beyond the enterprise boundaries by drawing stimulus and energy from a dynamic ecosystem of external stakeholders. Therefore, designing the IT of the future is not so much about designing structures and roles, but rather about creating and managing networks for innovation by defining motivations for engagement and modes of interaction, and designing open innovation processes and widespread change that are able to involve the entire organization, attract external expertise and stimulate talent and creativity in people.

4.2.5 A New Modality of Work

Together with the rise of business transformation, the IT department has become more and more central for the overall company performance and success. Today, IT is trying to push the organization to the so-called *fourth industrial revolution*. IT's current way of working is deeply different from what it was some years ago. As a result, IT is taking an evolutionary path that involves people, processes and technologies and aims at improving the organization's business performance. Taking the digital enterprise definition and translating it into IT responsibilities, we can say that IT has to consider four main components:

1. Strategic approach;
2. Digital Skills;
3. Customer-centricity;
4. Technology ecosystem.

Since IT holds a guidance position in the company's transformation, it is essential for the function to develop its own strategy that is coherent with the organization's business goals and priorities. The strategic approach that IT takes will determine the way in which the IT leaders will support their organization during the next years. To be sure that the strategy will be well executed, the IT department needs people that have digital competences that are supported by a culture that is lean, open and agile. As for the new digital enterprise, the client is at the center. The old way of considering processes, as being far away from the final customer, is gone. IT leaders realized that they can't ignore customers but rather they must identify and improve customer-facing processes and, from this, leverage the customer journey maps to alter and improve customer-facing processes (Robertson 2015). Indeed, the customers' ignorance can sometimes be transformed into a loss of possible new business opportunities.

Finally, the IT department is the company's interface with new technologies in the external environment. Therefore, it has to be aware of what is going on outside the company in terms of new technologies and trends. What is asked from the IT department is to stimulate the company' innovation by studying good practices and competition. Studying the technology ecosystem allows the department to create and stimulate new business opportunities and suggest new solutions to increase both operations and strategy.

The designing of an environment for digital innovation is not confined by the boundaries of an organization, as it requires a coherent re-thinking of sourcing methods and relationships with suppliers and partners.

No organization, however large, may believe itself to be autonomous in the face of the digital transformation challenge. Most of the stimuli and resources essential to innovation come from the external environment and it is from a creative interaction with the organization's boundaries that the best strengths for innovation are born.

Nowadays, the ecosystem in support of digital innovation is rapidly evolving and is destined, according to the findings of the Digital Transformation Academy of the Politecnico di Milano, to change substantially in the coming years. According to the CIOs of certain leading companies operating in Italy, the main sources of digital innovation external to the company have been, in order, *ICT Vendors, Customers* and *Consultancy Companies*. However, this reality is likely to change over the next few years with a decline in the role of *ICT Vendors* (−30%), *Consultancy Companies* (−15%) and *Outsourcers* (−38%) in favor of *Customers* (+10%) and, above all, of other stakeholders such as *Universities* and *Research Centres* (+27%) and especially *Start-ups* (+167%). The importance of imitating direct competitors also decreases (−33%) while the importance of imitating *Companies in other sectors* grows (+33%). A digital innovation ecosystem destined, therefore, to become increasingly open and dynamic, and which will require companies to call traditional sourcing policies into question.

Scouting for innovation, for example, must become broader and more dynamic. Relying only on the innovation roadmaps of the traditional vendors, in fact, does not seem to ensure access to digital innovation in the future. Companies must be able to monitor an offer in constant evolution and must have the courage to test unconsolidated solutions.

In this perspective, traditional procurement mechanisms are seen to be largely inadequate because they are incapable of opening up to the ecosystem's more innovative stakeholders.

In conclusion, the design of a new environment for digital innovation also involves an in-depth re-thinking of the methods used for recourse to the market, and agreeing to interact with new partners in accordance with open and dynamic logics that call into question traditional procurement and supplier management schemes.

4.3 Human Resource Department as a Transformation Facilitator

A McKinsey study (Bilefield 2016) affirms that 70% of organizations' transformations fail. The main reason for this is the employees' resistance to change followed by management behavior that does not support the change. In dealing with transformation, the organization is asked to deeply change the way in which it operates and the way in which it is organized. The problem is that the vast majority of people are change-averse and they don't see the potential behind it, so they resist it. As the numbers clearly show, people are the determining factor of an organizational success or failure.

As people so greatly influence the strategic results, the Human Resource department has become an important partner for the transformation success; HR employees are in a position to effectively motivate both the management and the organization's employees. Again, as with IT, support from the organization's board is crucial in exploiting this function's full potential. The Boston Consulting Group (Bhattacharya et al. 2016) defines it as an evolution in the work of the traditional HR. Indeed, they now have to make sure that the vision and values set in the beginning of the transformation journey are then translated into specific behavior and actions. In practical terms, they should work to ensure the creation of a new transformation company culture. Explaining to the employees what is the potential of a transformation and why the organization needs to change is a potential first step. If people feel involved in the transformation journey and they know the vision behind it, they will probably be more likely to adapt to it. In this way, the HR department can also contribute to the organization's overall innovation by promoting a risk-friendly culture in which people are not afraid of mistakes but are more interested in experimenting with different solutions and alternatives.

Ericsson—the multinational networking company that offers a wide range of services worldwide mainly related to telecommunications—is going through an important business transformation and is strongly relying on the HR office as a transformation facilitator. HR officer Bina Chaurasia, in an interview done by Mckinsey (2016), summarized the work done in the last years in three main points: (1) Single people strategy, (2) Integrated IT platform for HR and (3) Globalize HR processes. Ericsson's purpose was to align people with the business strategy and create an effective worldwide platform in order to increase employees' consciousness about the transformation journey and make them feel part of it.

4.4 Multi-Functional Teams

If we consider some of today's leaders in different markets such as Google, Spotify or Netflix, they have something in common: they have an agile way of working. As we said before, when ING Netherlands decided to transform its business model, they spent some days at Spotify. There, they learned to work in a new way which they referred to as agile. Today, ING Bank's fundamental work unit is the Squad. They defined it as a self-starting, autonomous team of up to 9 people who are responsible, end to end, for their own specific customer. They are built around different disciplines, different areas of expertise and different backgrounds.

Therefore, in order to succeed in the business transformation, people within the organization are no longer considered part of a specific department but rather of a specific project. In other words, the structure is redefined with different teams that are cross-functional and that work in different projects. In this way, people that have different knowledge and perceptions of the organization can interact and contribute. As a result, the traditional organization's barriers between different departments are removed and innovative thinking is stimulated on a regular basis. The coordination unit is the task force (Fig. 9). Adopting ING Bank terms, each Squad has people from different departments such as marketing, IT, Finance, Marketing and Human Resource. Every team has a Project owner that is responsible for the team's

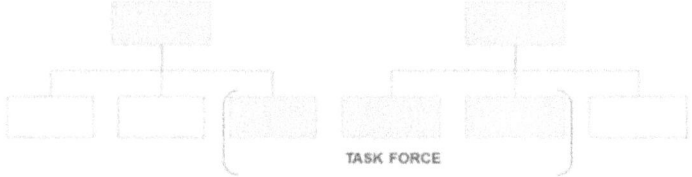

Fig. 9 Multi-functional teams

organization. He/she is the one that determines the duties and tasks that the people within the team have to do and the project priorities.

4.5 Coordinating the Transformation: The Evolving Role of the PMO

The Program Manager officer is the figure in the enterprise with the responsibility of centralizing and coordinating the different projects that a specific department has. He/She also has the duty of verifying that his/her function's project objectives, resources and deadlines are respected. Therefore, he/she is also the one in the department that has to coordinate different projects and be sure that they are all coherent with the department goals and dynamics. Traditionally, the PMO was located in a single department and was coordinating only that department' projects. Indeed, the PMO vision was more operative and specific to his/her department.

Since the business transformation is transforming the way in which the organization works and implements changes, the role of the PMO has evolved into a new, more strategic role: The Enterprise Program Manager Officer (EPMO). The EPMO is responsible for the organization and orchestration of the different projects at the enterprise level. The EPMO is more involved in the company core projects, and therefore has a more strategic approach to the business. He/She has a complete vision of the organization's strategic goals and projects and has to be sure that all the organization's main projects are respecting certain constraints. The EPMO projects have a higher impact on organization results for both operations and strategy. He/She has to be sure that the projects are coherent with the strategic guidelines, deadlines and resources. For this reason, he/she needs to have a deep knowledge of the organization's values and objectives and be able to manage and coordinate different projects at the same time. His/her role is crucial to succeeding in more important cross functional organizational projects, for example the transformation plan.

One organization that offers a practical example of this new role is the Italian company Amadori, which is facing an important organizational transformation. The IT department is becoming ever more central in the overall organization, which entails significant changes inside this function. Indeed, the CIO is responsible for many different business aspects: the traditional IT function, the new digital office, the business process transformation office and the organizational transformation plan implementation. Traditionally, the PMO was one of the members of the business process transformation office. She was in charge of the IT specific projects. More specifically, she was working to ensure that the different constraints of the different projects were respected and that they were all coherent with the IT department's objectives. However, in recent months, IT has become responsible for the organizational transformation plan. The transformation plan is made up of seven parallel streams that are all cross-functional and that involve different projects with different leaders. All the organizational departments are involved in the transformation plan and different departments are involved in each stream. As a consequence, multi-functional teams are working on it on a daily basis. Of course, the transformation plan is the most important thing for the company at the moment. For this reason, a person responsible for coordinating the different streams was needed. So, the IT PMO became the new EPMO in charge of coordinating and ensuring the success of the transformation plan projects. As one can understand, the EPMO role in this case is crucial for the success of the organization transformation plan. Without the best work orchestration and a clear vision of where the organization wants to arrive, the transformation plan could never be implemented.

For this reason, the EPMO is becoming a vital role for the organization. Each enterprise that is facing an important transformation requires excellent coordination along with a clear vision of the goals that have to be achieved through the different projects. As the organization grows and more and more cross-functional teams are generated, different people responsible for different teams will be needed. Above them, it is fundamental to have somebody who can orchestrate all their work and connect different teams.

5 Bring Innovation Inside the Organization

Since the digital era increased the speed at which organizations are asked to transform and reinvent themselves, they found new ways to innovate. They are required to keep on adapting to market changes and, if possible, anticipate these changes. For this reason, they have to go through a specific transformation journey that allows them to make the qualitative leap to survive the digital revolution.

Considering the transformation journey, the main output of this new enterprise re-organization and openness to the market is the enterprise-innovation (Fig. 10). In this configuration, the organizational structure and the new technologies available are smartly integrated in order to create disrupting innovation and gain a significant market advantage. Traditionally, innovation was owned by two main functions: Marketing and R&D. However, this configuration is no longer sufficient in light of the modern panorama.

Today, innovation must be done at a higher level, in which different areas interact in different moments, giving different perspectives and contributions.

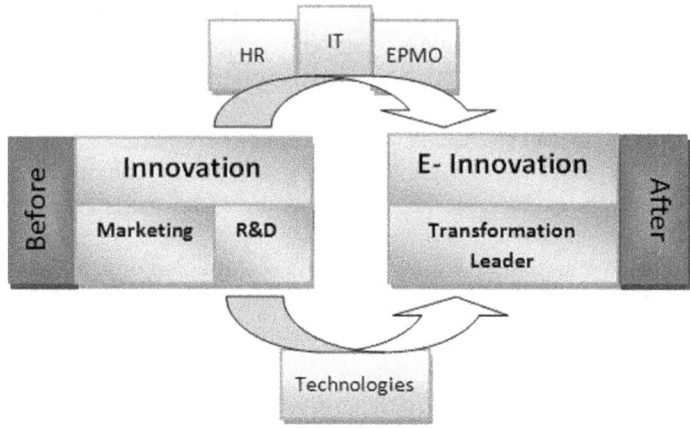

Fig. 10 From innovation to e-innovation

Indeed, the enterprise's innovation is the result of a deep cross—functional heterogenic analysis as well as the synthesis of many different business needs and ideas. It is a consequence of a new organization modality of work that considers the collaboration between different departments, together with a very strong consciousness of the external environment, and the most relevant strategic assets.

For this reason, innovation can't be driven and owned by only one or two functions but requires the contribution of many different functions and the central coordination of an organizational actor that can deeply understand both the market technologies and the new organizational configuration: the transformation leader. He/she takes the organization from a simple innovation to an enterprise innovation through a complex enterprise digitalization and re-organization by studying and exploiting the external resources and technologies.

References

Accenture (2015) Digital business era: stretch your boundaries, Report

Bhattacharya A, Bürkner H, Bijapurkar A (2016) What you need to know about globalization's radical new phase. BCG Perspectives. Available at https://www.bcgperspectives.com/content/articles/globalization-growth-what-need-know-globalization-radical-new-phase/. Accessed 20 Feb 2017

Bilefield J (2016) Digital transformation: The three steps to success. McKinsey & Company. Available at http://www.mckinsey.com/business-functions/digital-mckinsey/our-insights/digital-transformation-the-three-steps-to-success. Accessed 20 Feb 2017

Gartner (2014) Taming the digital dragon: The 2014 CIO agenda. Available at https://www.gartner.com/imagesrv/cio/pdf/cio_agenda_insights2014.pdf Accessed 20 Feb 2017

Gartner (2016) Gartner identifies seven best practices for an effective project management office. Improve project portfolio and program management practices and show value. Gartner, Stamford, Connecticut. Available at http://www.gartner.com/newsroom/id/3294017 [Accessed 8 Oct, 2016]

Govindarajan V (2016) Stop saying big companies can't innovate. Harvard Business Review. Available at https://hbr.org/2016/06/stop-saying-big-companies-cant-innovate Accessed 20 Feb 2017

Kane C, Palmer G, Philips D, Kiron A, Buckley N (2015) Strategy, not technology, drives digital transformation. MIT Sloan Management Review. Available at http://sloanreview.mit.edu/projects/strategy-drives-digital-transformation/ Accessed 20 Feb 2017

Nolan R, McFarlan F (2005) Information technology and the board of directors. Harvard Business Review. Available at https://hbr.org/2005/10/information-technology-and-the-board-of-directors Accessed 20 Feb 2017

Peterson R (2004) Crafting information technology. Information Systems Management

Politecnico di Milano (2016) Rapporti Osservatori 2016. School of Management. Available at: https://www.osservatori.net/it_it/pubblicazioni/rapporti

PwC (2016) Why are transformational leaders so important? Available at http://www.pwc.co.uk/services/human-resource-services/human-resource-consulting/why-are-transformational-leaders-so-important.html Accessed 20 Feb 2016

Rameshkumar S (2013) The Digital enterprise and the changing role of IT. tcs.com. Available at http://www.tcs.com/SiteCollectionDocuments/Perspectives/Digital-Enterprise-Changing-Role-IT-0613-1.pdf Accessed 20 Feb 2017

Robertson B (2015) Connecting process to customer: take the customer journey. Gartner.com. Available at https://www.gartner.com/doc/3168223/connecting-process-customer-customer-journey Accessed 20 Feb 2017

SAP (2016) Sapphire now (2016) Interview with Thomas Saueressig, CIO of SAP. Available at https://www.youtube.com/watch?v=UUAIf6M6U1I Accessed 20 Feb 2017

Wang R (2015) Disrupting digital business. Harvard Business Review Press, Boston

Westerman G, Bonnet D, McAfee A (2016) Leading digital. Harvard Business Review Press

Digital IT Governance

William DeLone, Demetrio Migliorati and Giovanni Vaia

Abstract This chapter presents how IT governance approaches are evolving to drive corporate transformation in this digital era. Today we are learning that new digital firms, embracing the digital transformation, are able to speed up and automate decision making processes, and build more agile, collaborative communities among internal resources, suppliers, customers and external experts. Consequently, the traditional view of IT governance may no longer be valid in today's digital enterprises. The question that arises from many scholars and practitioners is: to what extent do the well-established IT governance models still apply in the digital era? And, if they no longer apply, what new models and mechanisms can be proposed to address the changing demands placed on digital companies? This chapter reveals through the case of Banca Mediolanum how the traditional "functional" separation between business and IT is insufficient to support digital transformation. Digital initiatives must be well integrated into all organisational functions, as part of a unique, digital company DNA. Indeed, "Digital" Governance plays a critical role by supporting the change of organizational behaviours, pushing down digital decision-making, activating pervasive, horizontal, and collaborative communications, and supporting a shared decision-making culture.

The views expressed in the paper are those of the author and do not necessarily reflect those of the company.

W. DeLone
Kodog School of Business, American University, Washington D.C., USA
e-mail: wdelone@american.edu

D. Migliorati
Banca Mediolanum, Milan, Italy
e-mail: migliorati@mediolanum.it

G. Vaia (✉)
Ca' Foscari University, Venice, Italy
e-mail: g.vaia@unive.it

© Springer International Publishing AG 2018
G. Bongiorno et al. (eds.), *CIOs and the Digital Transformation*,
DOI 10.1007/978-3-319-31026-8_11

1 Introduction

In the past decade, digital technologies have substantially transformed the role of IT within firms (Westerman et al. 2011; McDonald et al. 2014; Gottlieb and Willmott 2014; Hirt and Willmott 2014). As companies progressively rely on mobile social media and cloud and big data, the very nature of the IT function within an organisation changes from providing reliable and cost-effective IT support to actively searching for new ways to leverage IT to create customer value.

The topic of IT governance has been widely discussed in different communities as a tool to multiply the benefits of IT in a business. The importance of past contributions notwithstanding, it stands to reason that a traditional understanding of IT governance might not adequately reflect the realities of a digital world. A more contemporary understanding of IT would need to take into account a number of recent developments; first, products and services are becoming increasingly more "digitalised," thereby blurring the boundaries between IT and business (e.g. marketing, sales, manufacturing) processes (Bharadwaj et al. 2013). Decision-making thus happens jointly in cross-functional teams and not by traditional, autonomous functional-level or bilateral decision-making. Second, the common way of thinking about IT as being subject to business authority is changings as IT becomes more business-aware and, consequently, more involved in "high-level" strategy-making. Finally, the high speed of technology development incentivises companies to develop governance arrangements deliberately aimed at simplifying and accelerating IT-related decision-making processes.

For scholars and practitioners alike, a question arises regarding the extent to which the well-established IT governance models still apply in the new digital context. Or else, if they no longer apply, what new models and mechanisms could address the changing demands placed on digital companies?

In this book, we state that new digital firms, in embracing the digital transformation, must accelerate and automate the decision-making processes (making them *quasi* real time decision making processes) and interactions, and build more agile relational paths among internal resources, suppliers, customers and external experts, to continuously improve the time to market and the capacity to introduce fast track innovation. Consequently, the traditional view of IT governance may no longer be valid in today's new digital enterprises.

Digital Governance (IT Governance in digital companies) plays a critical role in supporting the change of organisational behaviours, by pushing down digital decision-making, activating pervasive, horizontal, and collaborative communications, and supporting a shared decision-making culture.

2 Past, Present and Future Concepts

IT Governance

Although the academic and practitioner literature does not agree on a single definition of IT governance, many advocate a shared view that IT governance includes structural, process and outcome metric dimensions (Weill and Broadbent 1998; Bowen et al. 2007). According to this definition, IT governance delineates the roles and responsibilities for making IT-related decisions, designing effective decision-making processes and establishing performance assessment mechanisms.

Early research in the Information Science field has distinguished three broad IT-related decision categories—IT infrastructure, use and project management (Sambamurthy and Zmud 1999). Sambamurthy and Zmud (1999) proposed three major governance types, centralized, decentralized and hybrid, based on an organisation's IT-related authority patterns (Brown and Magill 1994).

Combining different perspectives, Wu et al (2015) delineate the necessary elements of an IT governance framework: "IT governance can be deployed via a mix of structures, processes, and relational mechanisms …Structures involve clearly defined roles and responsibilities and a set of IT/business committees such as IT steering committees and business strategy committees. Processes refer to formal processes of strategic decision making, planning, and monitoring for ensuring that IT policies are consistent with business needs. Finally, relational mechanisms, which include business/IT interaction and shared learning and communication, are crucial to the IT governance framework" (Wu et al. 2015, p. 502).

The influential study by Weill and Ross (2004), states that effective IT governance deploys:

- three different types of mechanisms: decision-making structures (such as committees, executive teams, and business/IT managers responsible for IT decisions), alignment processes (such as IT investment proposals and evaluations), and communication approaches (channels that spread principles and policies of IT governance and decision-making outcomes);
- five major IT decisions (IT principles, IT architecture, IT infrastructure strategies, business application needs, and IT investment and prioritization);
- three performance measures such as asset utilization, profit and revenue growth;
- six governance classifications available to IT organisations based on the ideal of political archetypes (Vaia and Carmel 2013). The Business Monarchy and IT Monarchy archetypes represent a centralized decision making structure; IT decisions are made by Chief Officers (CxOs) in the former and Corporate IT professionals in the latter. The Feudal archetype reflects a decentralized structure where business unit owners are the primary decision makers. The IT Duopoly archetype represents a two-party arrangement between a business group and IT executives. The Federal archetype functions as a "hybrid" decision making model and allows for creative business solutions within agreed-upon controls. Anarchy, where each small group can make decisions, is the sixth archetype (Fig. 1).

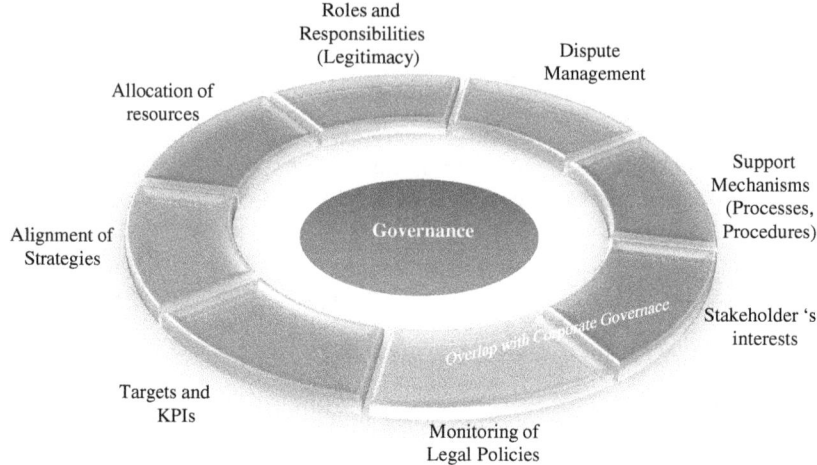

Fig. 1 Governance characteristics

Despite the importance of the IT Governance model proposed by Weill (2004), it shares several common traits with prior models; these traits significantly limited the applicability of these models in digital organisations. Namely, they overemphasize the role of hierarchy, propose robust structures that lack agility, and do not account for cross-functional synergies. Different studies in the IS field claim that effective governance and subsequent strategic alignment requires centralized governance structures (Sambamurthy and Zmud 1999; Kearns and Sabherwal 2007) and vertical communication (Martinsons and Davison 2007). That is, IT governance is characterized by an alignment at the top of the organisation, by vertical communication and a hierarchical culture, continuously searching to bridge the gap between IT and business.

This mechanistic approach to IT governance is inappropriate for firms today because Digital Transformation is much more than simply a transformation of technologies. The term Digital Transformation not only requires a new interaction between the technology and its user, but also a change in how people contribute to the creation of value and how a company organises its business.

Digital Transformation

Hirt and Willmott (2014) present an effective categorization of opportunities and consequences related to the digital transformation:

– Increasing pressure on margins and prices. Comparison between prices, in fact, has become easier through digital channels, particularly social media and the numerous websites aggregating different vendors' price information. This

particular factor is driving the market to a convergence in terms of prices and offerings, making competition fiercer;
- New competitors emerging from different industries. Digitalisation is removing entry-barriers and feeding product differentiation. Thus, new competitors can be represented by small start-ups, as well as by established players exploring new potentially revenue-generating businesses, such as Alphabet Inc. or Apple Inc., which are both stretching company boundaries (e.g. the Google Wallet or the future Apple Car). In the 2015 PwC Annual Global CEO Survey (PwC Italy 2015), 56% of the CEOs interviewed (728 CEOs out of 1300) responded that the some of their strategic moves for the next three years would be competing in new sector;
- Automation versus talent seeking. On the one hand, thanks to digitalisation, companies are more capable of automating processes, even in some more knowledge-intensive analytic areas (e.g. oncology diagnostics). This will inevitably increase the demand for data-literate human resources by large and medium-sized companies. On the other hand, there is an urgent need for digital talents who are able to use new technologies in areas where automation is not possible;
- Plug-and-play business models. The reduction in transaction costs due to digitalisation has provoked the disaggregation of value chains. In fact, nowadays third parties find it easier to provide their services to other companies in order to fill the gap that companies have in their chains;
- Worldwide standardization of demand and supply. There has been an increase in the number of systems that function across borders, of distribution on a global scale, and of a customer experience tending to uniformity;
- Continuous evolution of business models at higher velocity. Since the digital models continue to expand very quickly, companies must quickly adapt their models in order to satisfy the market's requests and continue to be profitable.

In other words, Digitalisation represents a cutting-edge re-organisation of the company's resources and customer relationships, as well as its products and services, with the ultimate aim of boosting revenues, improving efficiency and increasing the overall value of the company (McDonald et al. 2014). Furthermore, through new technology-enabling solutions, the digital transformation incorporates strategies and capabilities that change the rules of competition (McDonald et al. 2014).

Digital Governance

Governing this digital transformation requires the development of new abilities, new ways for managers to interact, and new mechanisms to generate innovation and support creative processes. Companies need to rethink IT governance in the context of the digital transformation. Leaders need to identify and resolve all the issues regarding the implementation of digital projects, and provide new polices, roles and responsibilities (Who is in charge? Who owns the digital processes? Who and what legitimize the allocation of responsibilities?).

"Digital" Governance should reflect these characteristics:

- *future proof*, adapt to continuously changing user needs, new technology adoption, and new markets. Governance is not a static framework based on company characteristics; rather, it is embedded in the constant flux of organisational re-design. Therefore, company history and culture shouldn't influence governance modes. Organisations need to be more liquid and attribute decision making responsibilities to employees who are closest to trends and customers.
- *cross-boundaries*, provide a frame for all digital initiatives. Digital governance needs to enable different perspectives across internal functions and external actors. The leadership must cross boundaries to empower employees to upset the traditional way that business is driven.
- *prone to innovation*, increase the pace of innovation and stay ahead of the competition. Digital governance should ensure that new technology investment decisions involve system users. Everyone in the organisation needs to be encouraged to work on relevant innovations, improve core processes, introduce new revenue sources, scout new products, search for new distribution channels etc. Digital governance offers a platform to balance incremental innovation and breakthrough projects, where risk taking and failure are encouraged. Innovation has been managed differently in many contexts, using top-down or bottom-up approaches. IBM or 3M have supported and used "idea killers" from the bottom to change the rules of competition. Others, such as the auto industry, have preferred to centralize the visioning and the development of new products and services. Governance today combines thorough rules and enhancing platforms (organisational and technological) and balances (top) management inspired innovation with the ability of the bottom to initiate front-end innovation.

To achieve and maintain these characteristics, digital governance needs to manage the following forces effectively over time.

2.1 The Power to Crowd

Modern technological advances are transforming the workplace. Crowdsourcing is one of the most disruptive phenomena that uses the power of collective intelligence combined with new digital opportunities. Crowdsourcing is built on the wide reach of the Internet, which connects a diverse group of individuals with a wide range of expertise, abilities, and problem-solving skills. "Crowds" can bring together more data, leading to a more accurate and intimate understanding of an environment. Researchers (Malone et al. 2010) have demonstrated how large, loosely organised groups of people can work together electronically in effective ways. They have shown how shared or group intelligence that appears in consensus decision making, like Wikipedia or Google, has already been proven to work.

Although online community users play the role of producers, innovators, and problems solvers, they are not part of established employee networks, as would be the case in traditional organisations. They are only temporarily connected to the organisation for a specific task or project.

Many digital companies are using hundreds of their own employees in crowdsourcing contests to encourage and challenge problem solvers to develop solutions from different perspectives, using a diversity of skills and knowledge. These workers experience increased enjoyment in the process of developing a solution to the contest challenge.

For example, in 2014, Banca Mediolanum, an Italian financial institution,[1] set up a collaborative platform hosting all the organisational projects that require multiple interactions between team members. An individual (called the sponsor), for instance, can launch a business challenge and everyone inside the Bank can contribute to the challenge and enrich it with new ideas. During the testing phase, the Idea Management project involved, on a voluntary basis, more than 600 employees, resulting in more than 20 challenges being activated and more than 170 ideas being shared. As reported by the Project Manager of the Idea Management platform "for twenty-five years, the Mediolanum Group has organised small groups of voluntary people focused on organisational improvement initiatives, but now the potential is incredible...something we could rename crowd-problem-solving."

2.2 Democracy

Digital media have the power to reconfigure the coordination of work across the organisation. By sustaining dialogue within and between organisations, digital media foster collaborative relationships and create transparency, connectivity, and sharing (Wollan et al. 2014).

Facebook, Twitter, LinkedIn, and smart mobiles, used by customers, employees and other company stakeholders are driving the cultural revolution promoted by digital. Today, we are fortunate to witness a spectacular transition towards democratic companies. Digital brings freedom and equality "in the pursuit of novelty and change" (Deschamps and Nelson 2014, p. 92).

One such change is the possibility for groups of employees coming from different functions to collaborate on a specific project, from beginning to end. The value of these groups, called "squads," is in their heterogeneity. Having people from different departments collaborating allows the team to consider elements from different point of views. This is the ING Bank approach. The company considers

---

[1] The bank was founded in 1997 as a "branchless" retail bank and, since its inception, has exclusively specialized in the provision of online financial services via multiple channels (telesales, Internet and mobile). During the period from 2007 to 2015, the bank pioneered a series of innovative banking solutions and, by 2015, had digitalised most of its client-facing activities and internal operations.

squads as a portion of a bigger entity that operates along similar lines. Squads involved in the same area of work are part of an overarching "tribe."

Shifting from a hierarchical/functional based organisation to a more liquid community based organisation requires a new type of coordination or governance.

Digital governance seeks to support democracy within a framework of rules. The concept of democracy is quite close to the "managed anarchy" at IBM. Rules and control mechanisms, in this scenario, guarantee individual participation in the community, and the opportunity to take part in the decision-making process.

This new perspective on organisational decision-making has potentially negative consequences. Personal interests and participation in all subjects could lead to rigidity, slowness, conflicts, and competence dilution, hence the importance of the aforementioned rules.

2.3 Inclusion

Another key motivating factor for undertaking a digital path is represented by the perceived need to strengthen the relationship with customers, suppliers and partners—the so-called stakeholders—and engage them in the activities of the company itself. Social media and new communication technologies in general, are opening cross-communication amongst these groups, allowing them to interact more freely and directly than companies and executives experienced in the past.

Customers give more prompt and direct feedback to companies, offering user-generated ratings and comparing products and prices. Companies, in turn, need to be as quick and direct as their customers. So, in order to establish a unique competitive advantage, producers must never stop engaging customers in the value creation process (Wollan et al. 2014).

Regarding the relationship with the other stakeholders, organisations today seem to be less aware of the potential of engaging their suppliers and partners. Currently, the dynamics occurring in the business-to-customer sphere, especially the need to establish long-lasting relationships and share knowledge, are underestimated. Indeed, more and more, organisations must digitally engage suppliers as well as integrate operations and the product development process.

Our research (Moretti et al. 2014) has shown how the adoption of a digital collaboration tool can mediate and support a more trust-based relationship between a client and his suppliers, and foster collaborative relationships that result in both operational and strategic innovation outcomes. We found that the exchange of high quality information and the use of effective communication tools are essential facilitators for process integration and for building trusting relationships in collaboration agreements. In our case study, client and supplier personnel perceived reciprocal professionalism, competencies and a willingness to share information to complete the task. In fact, the use of a digital collaboration tool changed relational governance in a short time, as trust between client and suppliers switched swiftly from affective attitudes to a more objective relation based on competencies.

2.4 Augmented Rationality

Organisations that experience positive returns on big data (1) pay attention to data flows as opposed to stocks, (2) rely on data scientists and product/ process developers rather than data analysts, (3) take analytics into core business, operational and production functions (Davenport et al. 2012).

As Davenport et al. claim (2012), "IT applications need to measure and report transparently on a wide variety of dimensions, including customer interactions, product usage, service actions and other dynamic measures. As big data evolves, the architecture will develop into an information ecosystem, a network of internal and external services continuously sharing information, optimizing decisions, communicating results and generating new insights for business."

The massive volume of data is changing technology infrastructures, competences and the IT organisation. Cloud technologies and virtual data marts, which allow data experts to use and share existing data sources—often not proprietary—without replicating them, enhance capabilities to effectively execute and automate real time decisions.

IT capabilities and architectures will evolve into an information ecosystem, based on a network of information stretched to provide support to managers, share performance results, and provide insight on business results, trends, changes.

3 New Models of IT Governance[2]

Digital transformation raises questions about the applicability of traditional governance approaches. As discussed above, companies need to make their IT department more business-aware, incentivize lateral communication and cross-functional learning, and promote the integration of previously disconnected functional units. Consequently, IT governance (Arkhipova et al. 2016), along with the models that describe it, need to evolve accordingly to account for the fundamental, digitally-enabled shifts.

In our research, we found three major digitally-driven organisational trends that appear to be driving IT governance changes within organisations: (1) horizontal communication, (2) democratic culture, (3) unified understanding between IT and business (see Table 1).

First, as both customer-facing and internal processes become more empowered by digital technology, the integration of multiple functional perspectives in developing new applications and processes has evidenced the need for increased transparency between different organisational units. The traditional models that

---

[2]This section has been partially published as a working paper at Department of Management, Università Ca' Foscari Venezia: *IT Governance in the Digital Era* by Arkhipova, Daria and Vaia, Giovanni and DeLone, William and Braghin, Carolina, July 2016, Working Paper No. 2016/12.

Table 1 Governance trends from traditional and digital perspectives

Traditional perspective	Digital perspective
Vertical communication	Horizontal communication
Hierarchical culture	Democratic culture
Shared understanding between IT and business	Unique understanding between IT and business

formally prescribed employees in different units to communicate through higher level representatives are no longer considered viable in a digital environment; governance models enabling smooth horizontal communication across peers appear to be more suitable.

Second, as business decision-makers become more IT-aware and vice versa, they start to demand a certain degree of autonomy in managing processes that may not directly fall into their domain of expertise. As a result, the hierarchical models that have historically envisaged a top-down line of command are now being perceived as ineffective as they preclude employees at the bottom of the hierarchy to make fast and informed decisions.

Finally, the theme that consistently appears in the literature suggests a blurring of the boundaries between IT and business, as any business process in a digital organisation becomes technology enabled and thus is indistinguishable from IT. In this regard, new "digital" IT governance models need to account for an ever-increasing overlap between the functional responsibilities of business and IT and a unified understanding that comes with it.

The transition from traditional to digital governance does not happen "overnight," however, and there will be temporal stages during which organisations will still have the vestiges of the old governance model co-existing with new digital governance elements.

As an organisation undergoes a transformation from traditional to digital, its governance systems are in perpetual flux. For a period of time (perhaps prolonged), a company will have decision-making processes that combine the elements of a legacy governance structure with new roles and mechanisms characteristic of digital organisations. In this regard, unlike fresh start-up companies that can build their digital governance structures anew, an established company has to accommodate both worlds, at least temporarily, until it can understand which IT governance model best suits its needs.

It is through experimenting that a company is able to understand to what extent new governance models are applicable in its specific organisational context. By subjecting a particular business unit to an experimental treatment, a company "simulates" a new governance model that remains operational in a particular domain and not in others. By testing new governance models, a company refines its approach to digital governance and prepares for rolling it out in other domains.

During an IT governance transition, governance types may be very idiosyncratic to each individual unit. Some units may require more autonomy, due to the nature of their work or their digital lifecycle; they will differ in the extent to which they are able to use and integrate third-party technologies in their operations. Projects in different domains will be managed differently. Units that are more adept at different technology platforms or have people that are more technology competent will push their own agenda and may bypass IT. Conversely, areas that are more dependent on IT and do not have technologies that could be easily integrated or used without local IT support, will continue to rely heavily on IT and governance changes will be less noticeable.

The type of governance structure adopted within an organisation will depend on the degree of its digital maturity. That is, as an organisation moves along its transformation path, its IT governance model undergoes corresponding changes, thus reflecting the requirements at each stage of the transformation. Based on our research, we propose a stage model of IT governance in which we theorize that governance models transition from traditional to digital through a series of six distinctive stages (Fig. 2).

Stage 1 represents a point of departure from the traditional governance models, such as Business Monarchy, that were widely adopted during the period preceding the digital revolution. This initial stage is characterized by clearly delineated roles and responsibilities between IT and Business, in which IT serves primarily as a service provider subordinate to the business. IT's involvement in high-level business-related decision-making is formally limited to occasional interactions.

At Stage 2, first steps towards collaboration between business and IT are being taken as IT becomes gradually involved in business-related decision-making

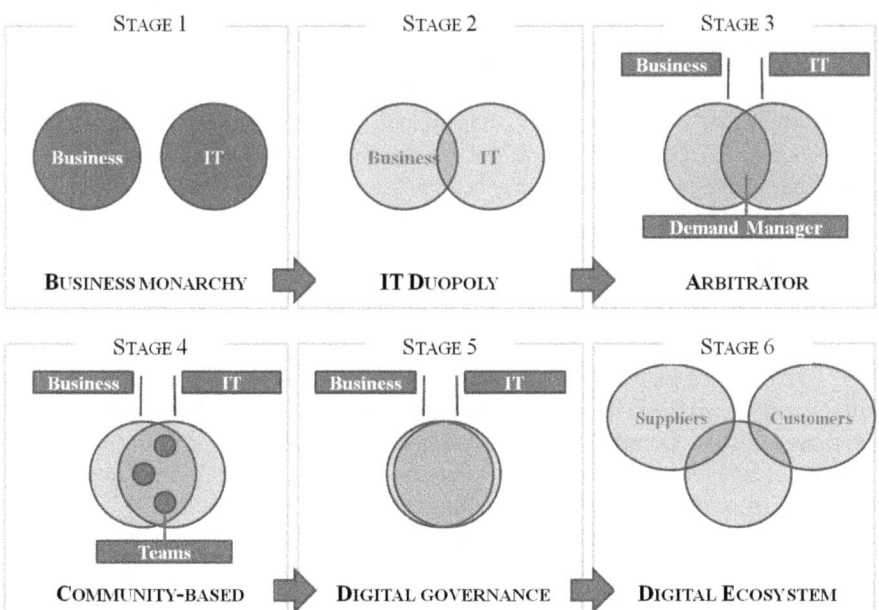

Fig. 2 Stage model of Digital IT Governance

processes and the two jointly manage technology projects. At this stage, IT and business interactions are designed on a bilateral basis so that IT becomes a key point of contact in technology-related communications for business units across the organisation, akin to the IT Duopoly model proposed by Weill et al.

Stage 3 is characterized by an increased volume of digital projects and growing overlap between IT and business. As the number and intensity of interactions between IT and business progressively increase, the governance model envisages a role of "Arbitrator" that is supposed to mediate this relationship and take pressure off IT and business in administering the company's digital project portfolio. Although it might involve an extra step in the decision-making process, the arbitrator's role is essential in taking a consolidated view of all digital activities taking place within an organisation, prioritizing and streamlining project workflow.

Stage 4 is defined by a gradual shift towards a more democratic mode of governance. The IT project organisation is team-based and the locus of decision making for most IT-related project issues moves down to the level of a single team; this mode is described as a Community-based governance model. Teams are composed of members with different functional backgrounds, thus fostering cross-functional communication and knowledge sharing.

At Stage 5, the notion of "business process" becomes synonymous with "digital process" and boundaries between IT and business are blurred as IT becomes entirely subsumed into the business process. Digital Governance blends its organisational units by eliminating *silo*-like work flows in favour of more transparent communication and knowledge sharing. The adoption of more collaborative work processes allows for the demonstration of a cohesive vision aimed at building an entirely new digital organisation in an orderly fashion.

Finally, we argue that there could be a point in time in which IT governance will span outside the traditional boundaries of an organisation and digital technologies will enable the involvement of consumers and suppliers in organisational decision-making. To that end, the governance becomes externally impacted and will be defined by the actions of the actors of a Digital Ecosystem.

4 Governing the Digital Transformation

Digital governance includes all those corporate mechanisms that allow coordinated actions and sharing of resources across organisational boundaries (Bonnet and Westerman 2014). Coordination across units would consist in aligning their multiple digital initiatives, while sharing refers to the use of common resources—such as technologies and people—and capabilities in order to enable digital change.

According to MIT and Capgemini Consulting research, governance represents one of the success factors in Digital Transformation (Tannou and Westerman 2012).

Digital transformation requires a balanced top-down/ bottom-up approach to motivate the coordination of various disaggregated digital investments. These represent the goal that an organisation should set in terms of the governance and leadership of the transformation.

As far as the mechanisms that can be used to make Digital Transformation work, three patterns have been identified: (1) the creation of shared digital units, (2) the arrangement of organisational-level committees, and (3) the establishment of new digital roles and relationships.

Shared digital units consist of independent units developing digital services for the entire company. For example, one of their responsibilities would be the development of needed digital technologies and services. These units would eliminate or, at least reduce, the redundancy of digital initiatives across the organisation, creating unique operations centres, such as an analytics competency centre, aimed at increasing the efficiency of digital efforts. Shared digital units are more agile; therefore experimentation is easier and innovation is more effectively stimulated. Another important responsibility of shared digital units is the design and development of company digital competencies necessary to overcome one of the most important obstacles to the transformation, the shortage of digital skill sets. As transformation processes require the right people, shared digital units would combine new human resources—experts in data analytics, social media, mobile technology and cloud computing—with existing staff, to create a balanced digital team. In addition, shared digital units would select employees from different business units for training and involvement in the transformation (Tannou and Westerman 2012).

Organisational-level committees represent another transformative governance mechanism. The established committees are primarily of two types: steering committees or innovation committees. Steering committees are in charge of determining the strategic and digital objectives, making investments, approving policies and defining priorities. Innovation committees, on the other hand, are more focused on a specific objective and concentrate on evaluating the business potential of emerging technologies. These committees have the critical goal of strengthening the relationships between the business and the IT department (Tannou and Westerman 2012).

New digital roles and relationships lead the Digital Transformation far beyond organisational structure. New roles might be established at the C-level, such as the Chief Digital Officer, who would be in charge of leading the digital units and aligning digital strategies from the top of the company with the requirements of the local units (Tannou and Westerman 2012). Others roles might be informal and focused on connecting digital units, e.g. the digital champions. They would be empowered based on their digital capabilities, attitudes and social roles inside an organisation, and would be effective at increasing employee engagement and commitment (Welch and McAfee 2013).

5 Causes of Failure and Related Remedies in Governing the Digital Transformation

Lack of urgency. Motivating people, giving them a goal to be accomplished urgently, is fundamental to a successful transformation process. If people are not motivated enough or not focused on the opportunity before them, they will not be persistent at carrying out their tasks and, thus, any effort to change would be in vain. Sometimes, leaders need to provoke a sense of fear about the present and the future of the firm if they want to obtain the desired reaction from management and employees, i.e. making them active participants in the organisational change. In other words, in some cases leaders need to make the current situation of the firm seem more dangerous than undertaking an unknown path, such as the one of transformation.

Failure to form a powerful group to guide the organisation transformation process. Involving employees and management is fundamental to their cooperation in a large-scale change. Nevertheless, the guiding coalition should certainly include influential leaders and managers and the full, active support of the CEO.

Absence of a strategic vision. A strategic vision consists of easy-to-communicate and emotionally appealing ideas drawing a picture of the future of the firm. The vision needs to clarify in which direction the organisation must move and the broad goals it needs to achieve in order to realise a successful transformation. The vision creates the destination and road map for the transformation. Without a strategic vision, a digital transformation is reduced to a simple list of confused and misaligned initiatives, resulting in the company missing its goals.

Failure to communicate the vision across the organisation. Without a credible communication plan, employees are not able to understand the reasons they should move from their comfort zone and commit to change. Moreover, communication by C-level executives must be consistent with their behaviours, because inconsistency between leader's words and actions are detrimental to employee buy-in.

A vision, to be powerful and to drive the transformation process, must to be communicated properly, using all possible available channels. Executives should not just talk about numbers and growth, but also about success stories and people in organisations that contribute to change in a positive way. If these messages are diffused effectively, employees will be more inspired to commit to the change and thus be engaged in the digital transformation. If such messages are passionate and transmitted from the guiding coalition to all colleagues, as peers, this could involve even more people on a voluntary basis, empowering the guiding group even more.

Failure to remove the barriers to reaching the new vision. Very often, the right communication of the vision itself is not sufficient to guarantee the desired results. It is necessary to remove the barriers that impede the realisation of the new vision. These barriers include the organisational structure and managers who decide not to commit to change. At the beginning of the transformation process, organisations need to eliminate the largest barriers, as they might undermine their digital path.

Missing short-terms goals. Without short-term goals to reach and celebrate, employees may become dissatisfied and the digital transformation will lose momentum. It is important to actively plan for and communicate short-term successes that are clearly linked to the strategic vision. In addition, the short-term horizons aligned to meeting the transformation goals will increase the expectations for managers, thus being a positive factor in a transformation effort.

Deviation from the final goal due to early apparent success. Declaration that the transformation has been achieved based on early success could be very dangerous, and could nullify all the change efforts. Digital transformation, in fact, implies a very long and complex process that requires years of work. Losing the right organisational tension and commitment to change would result in a failure to transform successfully.

A company must continue to promote new strategic initiatives, adapt to market changes and continuously innovate. Therefore, celebrating victories is particularly important for boosting employee engagement and for maintaining a sense of urgency throughout the organisation. However, employees must be reminded of the ultimate goals and vision.

Failure to institutionalize change into the organisational culture. Failure to change organisational culture could be one of the biggest detriments in the organisational transformation. In fact, no strategic change is to be considered complete if it is not incorporated in the company's daily activities. It is fundamental to root change into the organisational culture, celebrating the benefits of the new approaches adopted and making sure that the next generation of managers adopt and personify these new methods of working.

6 Governing Digital Transformation at Banca Madiolanum

Background and Role of IT & Innovation

Banca Mediolanum S.p.A. was established in 1997 in Basiglio (Milan, Italy) and today is part of the Mediolanum Group, founded in early 1982 by Ennio Doris, in partnership with Fininvest Group under the initial name Programma Italia S.p.A. The founders' initial idea was to create a financial group that was a bank, an insurance company, and a retail investment company.

Since the beginning, Banca Mediolanum proved to be an innovative omni-channel bank, leveraging information technology (digitalisation) to provide unique services and minimize costs. The bank initially employed 200 people with no branches in its distribution model, and therefore no fixed costs. At that time, customer management was handled via call-centres and television teletext, a rather advanced distribution channel at that time. In the early 2000s, Banca Mediolanum adopted the relatively unknown commercial use of the Internet and expanded its offerings with an on-line trading services platform called "My Trade." The Internet strengthened this innovative business model characterized by a multi-channel network without physical branches. For example, the bank became a frontrunner in home banking through extensive investments in its home banking service. In 2004, the Mediolanum Channel was born, representing the first Mediolanum Group's digital satellite TV channel, an evolution of the then-existing Corporate TV.

The first corporate web-site (www.mediolanum.com) was created in 2005; it serviced primarily the corporate and financial community. New services for the sales network were created, allowing Family Bankers to be updated on corporate news anytime, anywhere, by simply connecting with their laptops. In addition, the bank created a web-based vendor portal which enabled vendors to be connected with the bank and participate in on-line bidding. Through this portal, it was easier for the Group to manage and evaluate electronic offers in near real-time.

In 2007 the bank launched its first mobile service, B.Med Mobile, for what was then cutting-edge mobile phones, e.g. Nokia, HTC, and Blackberry. The bank introduced Interactive Voice Response (IVR), an automatic telephone response service active 24 h a day, in response to the requests of customers who wished to be more independent in executing their transactions, using consultants' support only when strictly necessary. Additionally, the bank introduced the electronic submission of official documents by e-mail, a further step forward in the digitalisation of its internal processes.

The bank created the Innova Portal of the Mediolanum Group, an intranet gateway containing all the technological tools, information, rules and procedures used both by call-centre operators and Family Bankers to retrieve information and deliver customer services. The gateway was also used by management and employees to work and "live" the organisation. The bank next introduced the Mediolanum Personal Marketing (later called Mediolanum Personal Branding Platform), basically a tool aimed at allowing the company to understand the best advertising method for each customer.

In the following years, the bank started to interact with customers on a global scale. New models of interaction and collaboration with customers were employed and digital relations started playing a key role in retail banking. During this period, the Mediolanum Group improved its training and learning area, inaugurating its Corporate University, defined as "a company inside a company," with the objective of training professionals to achieve excellent results in the relationships with customers. The Corporate University included the novel MedBrain, an on-line portal offering access to courses and documents for personal training.

The web-based platform B.MedNet was introduced in 2010, integrating Banca Mediolanum's four main areas—MedIntranet, Family Banker web-site, MedBrain and Corporate TV—and containing all the useful information and tools the sales network needed to operate at its best. The integration of the Corporate TV, allowed programs and videos to be broadcast directly on the portal and viewed from the sales network on any digital device.

From 2011 to 2013, the bank focused on two main streams: mobile apps and social networks. New apps were developed with a geo-positioning option able to find the location of the nearest ATM or Family Banker. Furthermore, the bank introduced on-line chat and the internet calling VOIP, with the benefit of multiple-calling which could connect three key stakeholders, the Banking Services Centre, the Family Banker, and the customer, getting higher efficiency and effectiveness in customer problem solving. Also, new mobile services were added, giving customers the ability to buy and sell government securities, to manage credit card accounts, to obtain information on life insurance policies and investments products, and to pay utility bills. The number of mobile transactions increased from 88.7 million in 2011 to 228 million in 2012.

Another important technological innovation introduced was the digital signature, which allows for more efficient processes while preserving legality. Since 2012, the digital signature extended to additional types of operations and products, increasing paperless procedures. The digitalisation of the subscription procedures has been successful, representing, in August 2015, 53.4% of total customers' subscriptions at the bank, Fig. 3 demonstrates the growth of digital subscription procedures.

Facebook, Twitter, YouTube and LinkedIn were introduced with specific editorial plans for each targeted audience. Indeed, social networking is a precious source of information and feedback from customers, but it also supports one of the Banca Mediolanum principles: support human relations. "Our company is a bank, developing its business starting from relations with people," reported the Social Media Manager, "it is a very simple concept, but critical for the bank success today." The Facebook fan page continues to be the most popular in the Italian banking sector with "fans" increasing by 77% since 2013 and registering a total number of 53,000 followers on Facebook.

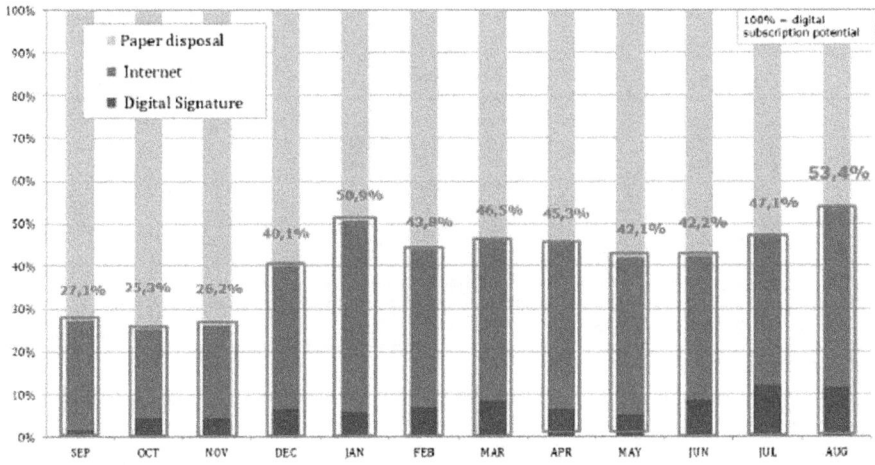

Fig. 3 Digitalisation of the subscription procedures from 2014 to 2015

The Digital Transformation Era

In this context, Banca Mediolanum started its comprehensive Digital Transformation program in 2013. The bank gave birth to a broader unified organisational plan, supported by a true digital vision. From being the main driver of commercial innovations at the bank, Information Technology became a tool for Banca Mediolanum employees to improve internal communications, collaborations and a sense of belonging to the organisation. The real innovation of the digital era is therefore the shift from technology as a stand-alone tool to an enabler of collaboration and relations among people, thusly augmenting human assets.

Digital initiatives launched since September 2013 were: the Digital Workplace program, the Collaborative Improvement, encompassed in the Idea Management project, the Knowledge Management, and the Digital School, as well as a research and monitoring unit on the Digital Transformation. Figure 4 summarizes the Bank's digital initiatives.

The Digital School was the tool through which the bank motivated employees with specific and timely goals, to be reached during the transformation process.

	Name	Description	Digital Workplace projects
1st area	Communication	Information sharing tools and transparency • Favouring transparency of corporate information • Redefining internal communication flows to overcome the use of e-mails • Enhancing corporate communication and the sharing of people's roles and responsibilities	Corporate news and comments ("Following"), Content access ("Inform myself")
2nd area	Collaboration	Environments for carrying out interactive activities, listening and talking with people • Redefining processes and internal project management • Sharing of the advancement in a project • Suggesting experts for problems resolution • Knowing priorities and activities advancement • Sharing of experience and colleagues best practice	Idea Management "Let's collaborate" programmes
3rd area	Activities	Self-service applications and end-to-end services, management of the business processes and data • Enhancement of the performance • Processes automation • Work and mobility	To-do list ("Organise myself")
4th area	Knowledge	Repository and content research, capitalisation of people and organisational know-how • Enhancing research and access to internal know-how • Access to people competences • Management and sharing of specialised information and issues • Training and giving assistance to people	Knowledge Management

Fig. 4 Digital initiatives since 2013 at Banca Mediolanum

This large scale engagement,[3] which included influential leaders, supported the dissemination of the digital strategic vision and clarified the direction in which the organisation was moving. People were therefore engaged and committed to a concrete road map. Moreover, in 2014, 30% of the digital budget was devoted to communication and advertising campaigns, demonstrating the great emphasis that Banca Mediolanum placed on its digital transformation.

Two additional important organisational projects started in 2014: the SMART Program and the Digital Workplace project. The SMART Program aimed at defining innovative and transformational initiatives for the bank's operating model. The Digital Workplace project, on the other hand, was born with the intent of driving the organisation towards more open and transparent forms of collaboration capable of disrupting organisational silos, in order to reach a more proactive approach to customer stimulus and knowledge sharing. According to the Digital Workplace Senior Manager, from the beginning, the main goal of the project was "[...] increasing the level of trust between people, in order to trigger a mechanism of mass collaboration and, thereby, enhancing the organisational performance." The purpose of the Digital Workplace is to create a unique digital space through which people can work better, having at their disposal all the information they need for carrying out their daily activities and through which they can collaborate and share information, ideas, and documents. This electronic collaborative tool helped to remove barriers to the successful implementation of the transformation.

The Digital Workplace is the natural evolution of the intranet platform of the company, integrating all the Web 2.0 functionalities—e.g. content creation tools, comments, status updates, tagging, etc.—with the Enterprise Resource Planning (ERP) platform—sending notifications and tasks to people based on their project role or activities. This active planning and monitoring was a powerful tool to increase expectations and maintain the organisational tension to change.

Subsequently, the bank realized that the tension to change needed a boost in order to engage more people while celebrating victories, leaders, and champions of the transformation. Therefore, the change management team conducted an Organisational Network Analysis (ONA)[4] aimed at amplifying the Digital Transformation process and discovering those who, inside each network, have the characteristics to sustain the transformation. These key change agents are also known as pivots or digital champions and are named "vertices" or "nodes" in graph theory (the area of mathematics that formalizes the study of models to allow pairwise relations between objects). Specifically, the analysis consisted in asking 2044 users four specific questions related to their relationships with other people inside the company. The output was a list of digital champions to involve in the design and leadership of digital projects. Starting with these results, the digital

---

[3]In 2013 the catalogue of on-line courses doubled in number, with an increase of 164% for courses related to digital projects. As a consequence, the total number of participants increased by 71% from 2012.

[4]ONA is a quantitative technique for studying and graphing the relations within an organisational network.

transformation team set the groundwork for the unified company platform: the Digital Workplace.

The two main initiatives contained in the Digital Workplace project are represented by the Idea Management and by the Knowledge Management projects. While the former was initiated in 2014, the latter started in 2015.

As presented above, the Idea Management project consists in a collaborative improvement platform (both on-line and off-line) that can host all the organisational projects that require multiple interactions between team members. An individual, also called the sponsor, can launch a business challenge and potentially everyone inside Banca Mediolanum can contribute to the challenge and enrich it with new ideas. These challenges can generate a real organisational project, with human, technological and economic resources. The Idea Management project started in 2014 with a "pilot" version involving 83 pivots (or digital champions) on a voluntary-basis and then continued in 2015 with the involvement of more than 600 pivots, covering 60% of employees across organisational departments. These employees experimented with innovative techniques for improving collaboration, proposing projects, increasing operational efficiency and organisational initiatives, and using advanced technological tools. This resulted in the activation of more than 20 challenges, the sharing of more than 170 ideas and the generation of 3000 associated comments. Moreover, more than 20 projects were launched based on positive cost-benefit analyses.

As reported by the Project Manager of the Idea Management, "…the concept at the base of the Idea Management it's the same… but in this case the potential is incredible because it is not only about involving small groups, but rather, the entire organisation." Therefore, the long-term benefits of the Idea Management platform in terms of augmented intelligence are impressive.

The second most important digital project in 2015 was the Knowledge Management platform. The Knowledge Management platform at Banca Mediolanum consists of a digital platform serving as a unique organisational knowledge repository, which is useful to both the front-office and the back-office, as well as to the sales network and anyone needing access to the company's information and data. The platform is built in Office 365®, which has been integrated with other tools and Microsoft applications—e.g. e-mails, Skype®, Word®, Excel®, etc.—linked to a database. Banca Mediolanum's management expects that, by the end of 2016 the project will be generating savings of about 600.000 € per year.

Taking into consideration the bank's many digital projects, the HR Senior Manager reports that, "the use of technological assets has differential elements representing one of the main pillars on which the business of Banca Mediolanum is based, thus being part of its DNA."

The bank, indeed, was born with the idea of exploiting cutting-edge technologies and putting them at the service of customers' in every given technological "era." These technologies, however, provide a value in establishing and managing the relationships the bank has with its stakeholders (customers and employees primarily). Human relations have thus been shaped with the use of digital technology as enabler and facilitator. Another motivation for the bank to capitalise its efforts towards Digital Transformation arises from the necessity to maintain and enhance its position in the market. Nowadays, the bank is performing well, not only economically but also operationally, and it is determined to achieve even higher levels of responsiveness, flexibility and speed. These improvements must be reflected in better customer services and products, which are the real focus of Banca Mediolanum. "As Ennio Doris says", reported the Project Manager of the Idea Management, "any firm that sits on its current wealth will be wiped off by the next generation of companies," and he continued, "[...] because those who will survive, will be the ones that would be able to ride this Digital Transformation wave."

All these digital-driven changes have provided the bank with the following benefits (Thomas et al. 2014):

- the ability to be creative, test various ideas and, through the support of the new digital tools, see what works and what does not in real-time, avoiding recurrent errors;
- the ability to learn, i.e. retain knowledge acquired through experiments and a propensity for self-learning;
- the ability to judge, i.e. to balance the facts, the potential future drawbacks, risks and opportunities in order to make informed decisions;
- the ability to cooperate with other bank employees intensively, in order to create new value.

Therefore, according to the HR Senior Manager, "the future benefits are expected to be perfectly aligned with the ones that we are today generating and on which we are currently working ...so the main goal is to change our mind-set and making today's approach at Banca Mediolanum an end-to-end approach."

From our perspective, Digital Transformation should be considered more as a transformation of people's behaviours and the organisational culture rather than just a technological change. Technology is in fact often defined as an *enabler*, i.e. an instrument used to make things happen. This aspect was clearly emphasised by Banca Mediolanum's *Innovation Senior Manager*:

> The main motivation that spurred Banca Mediolanum to undertake the Digital Transformation path has been the one of creating a company that would reflect the new way of working together, built through a change in people's behaviour, as well as through the use of new technologies. These new behaviours would imply a set of new rules involving the collective intelligence, the mass collaboration.

However, all the projects and activities carried out from 2013 to 2016 were implemented on the old governance structure, creating a sort of organisational magma. No new deliberated organisational forms and mechanisms have been implemented during the transition. Thus, we are now observing new emergent forms that will be consolidated and institutionalized in the following years.

Indeed, the IT department is assuming new roles, essential for supporting the digital transformation. In some cases, it is the CIO who drives the digital initiatives, whereas in other cases, ad hoc teams, formed by both IT and business personnel, are in charge of drawing up the digital agenda (Westerman et al. 2011). However, it is not just IT's role that must change and adapt to technological and organisational transformations, but also the structure of the IT function, which needs to be reshaped in order to allow the company to be more reactive and respond to market challenges.

Governance leverages the change triggered in people's behaviour and the culture by funnelling the digital efforts towards a more structured organisational transformation.

Banca Mediolanum is using *digital* to bring about a cultural and social transformation first, recognising how a comprehensive change will only be realized by engaging people and pushing them to take actions towards change. For the bank, this requires blending its organisational units by eliminating *silo*-like workflows in favour of a more transparent, community based organisation. The adoption of more collaborative ways of working will allow for the broadcasting of a cohesive vision aimed at building an entirely new organisation in an orderly fashion.

Banca Mediolanum's digital transformation would not have been possible without parallel changes in digital governance. Digital IT Governance at Banca Mediolanum has become more liquid and more widespread across the organisation. For example, HR managers are leading IT teams, local business units are empowered to use and maintain information technology, and IT personnel are disseminated across many corporate functions. Today, IT is melded into the bank's functions and processes. The Banca Mediolanum case demonstrates the rise of new IT governance models, including integration of governance roles and a prototype of a community-based governance. In the future, we expect that the bank will adopt a governance model that spans beyond the traditional boundaries of the company, enabling the participation of external actors, such as consumers and suppliers, in the Bank's IT decision-making, resulting in the creation and development of a Digital Ecosystem.

The bank realized that digital transformation, innovation and the evolution of new governance modes and mechanisms must move together. The Bank has learned that the implementation of new governance modes is critical to the success of the digital transformation journey. Naturally, this journey doesn't happen overnight (especially for a bank!).

Until the early 2000s, Banca Madiolanum adopted a traditional governance approach, i.e. a *Business Monarchy*. During that period, IT was a service provider subordinate to the business, and limited to technological evaluations. A management committee—General Management Project Committee—composed of the bank's General Manager and C-level executives ensured the alignment of business goals with IT projects. Once the project was approved, the members of the investment committee, i.e. the General Manager and the CEO, jointly estimated financial, human and technology resources required for successful project completion. As the Bank moved toward a digital transformation phase, an *IT Duopoly* governance model was adopted in which the business and the IT department created a liaison for business software projects and an IT representative was assigned to its management. While technical development, testing and deployment were the responsibility of IT, most of the choices related to software business requirements and functionality were made jointly with each business unit. Briefly, the pre-digital stage was characterized by a collaboration of IT specialists and business users on software business requirements and functionality.

Since the Bank's digital transformation required a rapid increase in IT projects (especially from 2013 forward), the Bank created a governance structure that supported continuous application delivery. To ensure rapid development, the Bank created the integration role of Demand Manager to approve and manage the growing number of digital projects. The Demand Manager has the responsibility to gather and collect information related to each individual digital project and to evaluate the strategic alignment of the project to the enterprise strategy. This governance approach has helped to identify synergies between different projects and to mediate the relationship between IT and business. The Demand Manager defines business requirements, evaluates technical features, and supports the Project Committee in its decision-making. Navigating toward a new organisational form, the independent "arbitrator," mediates between the business and IT culture and priorities and is the initial new digital governance form of future digital enterprises, where widespread technologies are fully integrated and fused in all corporate units.

Eventually, continued digital transformation will require heterogeneous agile teams that make governance boundaries between business and IT more blurred and overlapping. For instance, at Banca Mediolanum, a community-based governance structure drove the Idea Management project. Cross-functional, self-regulating community members were involved in application planning and development. These community members were responsible for budget spending, technology choices, and services to be delivered. So, the IT decision-making is not determined by the boundaries of a single business unit or IT function but rather, it involves extensive, transparent, lateral communications with peers across the organisation. The bank envisions a digital governance organisation where developers, technicians and functional employees take responsibility for their specific parts of a project and a strong hierarchical control is no longer needed.

To conclude, Banca Mediolanum has enhanced IT's role in business-related decision-making across the entire organisation. IT has become a central element of the corporate governance system that takes both a ground level and consolidated view on all digital activities. This approach challenges the core logic of the traditional governance models, revealing new governance models under the umbrella term *"Digital Governance."*

7 Concluding Remarks

Banca Mediolanum's case demonstrates that the evolution of an organisation's culture and governance models is critical in order for a company to benefit from the digital revolution. Indeed, the Bank's digitisation is the most dramatic and irreversible corporate transformation in recent decades. The lesson learned from the Bank's experience is that mechanistic, hierarchical governance structures are inadequate in a digital enterprise, where values and mechanisms such as democracy, inclusion, and continuous engagement across different digital communities are fundamental.

Accordingly, staff and management interactions have changed in order to remove the barriers that interfere with motivating progressive groups and their ability to guide an organisation's transformation process and to accomplish goals quickly. Yes, speed is critical now for two reasons. First, the rapid introduction of new disruptive technologies has changed the perception of time in society: we can grow, learn and live faster than in the past and we embrace this change. Second, digital companies such as Uber or Airbnb can exploit the value of information technology and rapidly organise and create new business models. This fast track to success is attractive to companies. To quickly exploit the value of technological advances, company employees must share a common vision, the same short-term goals and be acclimated to continuous change (we don't know if and when a new steady state will be reached, so continuous change is the new norm).

The traditional "functional" separation between business and IT is detrimental today. Digital initiatives must be well integrated into all organisational functions, as part of a unique, digital company DNA. After initial efforts to bring the business and IT closer together through integration roles such as Banca Mediolanum's Demand Manager, companies should move toward greater overlaps between business and IT, until digital competences are widespread in all business functions. Actions that can facilitate an organisation's digital transformation include (1) steadily moving IT professionals across the organisation and (2) creating small IT units in other functions. The result is the creation of heterogeneous digital communities of peers. After creating digital communities within the organisation, the next step toward becoming a digital organisation is the addition of external actors, including customers and suppliers, within these digital governance groups and communities.

The implementation of a digital enterprise calls for a "behavioural shift" that recognizes internal digital champions and digital customers and creates a seamless digital experience for all stakeholders. The active involvement of the Human Resources (HR) function is critical to this shift. The HR function should define the engagement action plan, incentives for participation and the strategy to propagate the digital transformation across the entire enterprise. The launch of a training and education program through a dedicated organisational unit is key to creating compelling content and leveraging dissemination dynamics in order to help employees (1) adopt new digital communication methods (beyond e-mail), (2) manage communication overflows (for instance educating people to correctly perform tagging and document archiving) and (3) use social-media strategically.

As a first step toward digital governance, managers should create an employee experience of "being connected" by organising information, applications and services into one single environment with easy end-user access. This internal digital experience also requires new corporate communication services, such as tweet-walls and blogs as well as a social knowledge platform (a Knowledge Management System 2.0). Mobile access to information services, both inside and outside the company, is fundamental to the launch of a company's first digital communities, with their own social profiles, policies, and facilitation roles. With these technology advances, employees will feel comfortable with the informal and socially oriented communication styles, which encourage learning.

As a next step toward digital governance an organisation encourages digital protagonists! Search for digital champions with high potential and enable full collaboration among customers, the sales network and the company social network by extending multimedia communication features to all. Personal objectives, KPIs and support from a 2.0 intelligence system are vital to the cultivation of digital talent. Digital talent across the enterprise is the foundation of a truly shared and democratic digital governance program capable of yielding timely digital innovations. Full digitalisation and socialisation of strategic internal and external processes will then result in the optimization of operations and enable digital communities to innovate company products and services, thereby achieving an impactful digital transformation.

References

Arkhipova D, Vaia G, DeLone W, Braghin C (2016) IT Governance in the Digital Era, Ca' Foscari Working Paper No. 2016/12
Bharadwaj A, El Sawy OA, Pavlou PA, Venkatraman N (2013) Digital business strategy: Toward a next generation of insights. MIS Q 37(2):471–482
Bonnet D, Westerman G (2014) We need better managers, not more technocrats. Harvard Bus Rev
Bowen PL, Cheung M-YD, Rohde FH (2007) Enhancing IT governance practices: a model and case study of an organization's efforts. Int J Account Inform Syst 8(3):191–221
Brown CV, Magill SL (1994) Alignment of the IS functions with the enterprise: toward a model of antecedents. MIS Q 18(4):371–404
Davenport TH, Barth P, Bean R (2012) How 'big data' is different. MIT Sloan Manage Rev 54:1
Deschamps JP, Nelson B (2014) Innovation governance: how top management organizes and mobilizes for innovation. John Wiley & Sons
Gottlieb J, Willmott P (2014) The digital tipping point: McKinsey Global Survey Results. McKinsey & Company, McKinsey Quarterly
Hirt M, Willmott P (2014) Strategic principles for competing in the digital age. McKinsey & Company McKinsey Quarterly. Retrieved from McKinsey Insights & Publications
Kearns GS, Sabherwal R (2007) Strategic alignment between business and information technology: a knowledge- based view of behaviors, outcome, and consequences. J Manage Inf Syst 23(3):129–162
Malone TW, Laubacher R, Dellarocas CN (2010) The collective intelligence genome. MIT Sloan Review, 51(3)
Martinsons M, Davison R (2007) Culture's consequences for it application and business process change: a research agenda. Int J Internet Enterp Manage 5(2):158–177
McDonald M, McManus R, Henneborn L (2014) Digital double-down: How far will leaders leap ahead? Accenture Strategy
Moretti A, Vaia G, DeLone W (2014) Reframing outsourcing through social networks: evidence from Infocert's case study in working paper series vol. 6
PwC Italy (2015) Competing in a marketplace without boundaries. PwC, 18th Annual Global CEO Survey Italy
Sambamurthy V, Zmud RW (1999) Arrangements for information technology governance: a theory of multiple contingencies. MIS Q 23(2):261–291

Tannou M, Westerman G (2012) Governance: a central component of successful digital transformation. MIT-CDB and Capgemini Consulting Joint Research Program on Digital Transformation

Thomas RJ, Kass A, Davarzani L (2014) From looking digital to being digital: the impact of technology on the future of work. Accenture Institute for High Performance and Accenture Technology Labs. Retrieved from http://www.accenture.com/sitecollectiondocuments/pdf/accenture-impact-of-technology-april-2014.pdf

Vaia G, Carmel E (2013) Reshaping the IT governance in Octo Telematics to gain IT–business alignment. J Inf Tech Teach Cases 1:1–8

Weill P, Broadbent M (1998) Leveraging the new infrastructure: how market leaders capitalize on information technology. Harvard Business School Press, Boston

Weill P (2004) Don't just lead, govern: how top-performing firms govern IT. MIS Quarterly Executive 3(1):1–17

Weill P, Ross J (2004) IT governance: how top managers manage IT decision rights for superior results. Harvard Business School Press, Boston

Welch M, McAfee A (2013) Being digital: engaging the organization to accelerate Digital Transformation. MIT-CDB and Capgemini Consulting. Retrieved from https://www.capgemini-consulting.com/resource-file-117access/resource/pdf/being_digital_engaging_the_organization_to_accelerate_digital_transformation.pdf

Westerman G, Calméjane C, Bonnet D, Ferraris P, McAfee A (2011) Digital Transformation: a roadmap for billion-dollar organizations. MIT-CDB and Capgemini Consulting

Wollan R, Palmer D, Jain N (2014) Digital customer: it's time to play to win and stop playing to lose. Accenture, 2013 Global Consumer Pulse Research. Retrieved from https://www.accenture.com/us-en/~/media/Accenture/Conversion-Assets/DotCom/Documents/Global/PDF/Strategy_3/Accenture-Global-Consumer-Pulse-Research-Study-2013.pdf#zoom=50

Wu SPJ, Straub DW, Liang TP (2015) How information technology governance mechanisms and strategic alignment influence organizational performance: insights from a matched survey of business and it managers. MIS Q 39(2):497–518

GPSR Compliance

The European Union's (EU) General Product Safety Regulation (GPSR) is a set of rules that requires consumer products to be safe and our obligations to ensure this.

If you have any concerns about our products, you can contact us on

ProductSafety@springernature.com

In case Publisher is established outside the EU, the EU authorized representative is:

Springer Nature Customer Service Center GmbH
Europaplatz 3
69115 Heidelberg, Germany